# 75th Annual Meeting of the Labor and Employment Relations Association and LERA@ASSA Meeting 2023



Labor and Employment Relations Association Series

Detroit, Michigan, USA
1 - 4 June 2023

New Orleans, Louisiana, USA
6 - 8 January 2023

**Editor:**

**J. Ryan Lamare**

ISBN: 979-8-3313-0006-7

**Printed from e-media with permission by:**

Curran Associates, Inc.
57 Morehouse Lane
Red Hook, NY 12571

**Some format issues inherent in the e-media version may also appear in this print version.**



Printed with permission by Curran Associates, Inc. (2025)

For permission requests, please contact Labor and Employment Relations Association (LERA) at the address below.

Labor and Employment Relations Association (LERA)
University of Illinois at Urbana-Champaign
121 Labor and Employment Relations Bldg.,
504 E. Armory Ave.
Champaign IL 61820
USA

Phone: (217) 333-0072
Fax: (217) 265-5130

LERAoffice@illinois.edu

**Additional copies of this publication are available from:**

Curran Associates, Inc.
57 Morehouse Lane
Red Hook, NY 12571 USA
Phone: 845-758-0400
Fax: 845-758-2633
Email: curran@proceedings.com
Web: www.proceedings.com

LABOR AND EMPLOYMENT RELATIONS ASSOCIATION SERIES

# Proceedings of the LERA 2023 Meetings

---

LERA@ASSA Meeting
January 6-8, 2023
New Orleans, LA

LERA 75th Annual Meeting
June 1-4, 2023
Detroit, Mi

LABOR AND EMPLOYMENT RELATIONS ASSOCIATION SERIES
(Formerly the Industrial Relations Research Association)

*Proceedings of the Annual Meeting*
Annual Research Volume
*Perspectives on Work* Magazine

Inquiries and other communications regarding membership, meetings, publications, and general affairs of the Association, as well as notice of address changes, should be addressed to the LERA office.

LABOR AND EMPLOYMENT RELATIONS ASSOCIATION
University of Illinois at Urbana-Champaign School of Labor and Employment Relations
121 Labor and Employment Relations Building, MC-504
504 East Armory Avenue, Champaign, IL 61820
Telephone: 217/333-0072 Fax: 217/265-5130 E-mail: LERAoffice@illinois.edu
Website: http://lera.memberclicks.net

## OFFICERS OF LERA FOR 2022-2023

**PRESIDENT:**
Paul Clark, Pennsylvania State University

**PRESIDENT ELECT:**
William E. Spriggs, AFL-CIO

**PAST PRESIDENT:**
Wilma Liebman, Former Chair NLRB

**SECRETARY-TREASURER:**
Andrew Weaver, University of Illinois Urbana-Champaign

**EDITOR-IN-CHIEF:**
J. Ryan Lamare, University of Illinois Urbana-Champaign

**CHAPTER ADVISORY CHAIR:**
William Canak, Middle Tennessee State University

**EXECUTIVE DIRECTOR:**
Emily Smith, University of Illinois Urbana-Champaign

**REGIONAL VICE PRESIDENTS:**
Jim Pruitt, Kaiser Permanente (West Region)
Beverly E. Harrison, Arbitrator/Mediator (East Region)
Robert Chiaravalli, Strategic Labor and HR LLC (Mid Region)

**EXECUTIVE BOARD MEMBERS:**
Peter Berg, Michigan State University
Julie Farb, AFL-CIO
Janet Gillman, Oregon Employment Relations Board
Shannon Gleeson, Cornell University
Quinton Herbert, City of Baltimore
Tamara Lee, Rutgers University
Kevin Legel, Ford Motor Company
Glenard Middleton, AFSCME Maryland Council 67
Deborah Moore-Carter, City of Baltimore
Deborah Mueller, CSEA Local 1000, AFSCME
Dionne Pohler, University of Toronto
Javier Ramirez, Federal Mediation and Conciliation Service
Christine Riordan, University of Illinois Urbana-Champaign
Sean Rogers, University of Rhode Island
Marc Weinstein, Florida International University
Christy Yoshitomi, Federal Mediation and Conciliation Service

**LEGAL COUNSEL:**
Steven B. Rynecki, von Briesen, Purtell & Roper, S.C.

## PAST PRESIDENTS OF LERA:

1948—Edwin E. Witte
1949—Sumner H. Slichter
1950—George W. Taylor
1951—William Leiserson
1952—J. Douglas Brown
1953—Ewan Clague
1954—Clark Kerr
1955—Lloyd G. Reynolds
1956—Richard A. Lester
1957—Dale Yoder
1958—E. Wight Bakke
1959—William Haber
1960—John T. Dunlop
1961—Philip Taft
1962—Charles A. Myers
1963—William F. Whyte
1964—Solomon Barkin
1965—Edwin Young
1966—Arthur M. Ross
1967—Neil Chamberlain
1968—George P. Shultz
1969—Frederick Harbison
1970—Douglass V. Brown
1971—George Hildebrand
1972—Benjamin Aaron
1973—Douglas H. Soutar
1974—Nathaniel Goldfinger
1975—Gerald G. Somers
1976—Irving Bernstein
1977—F. Ray Marshall
1978—Charles C. Killingsworth
1979—Jerome M. Rosow
1980—Jack Barbash
1981—Rudolph A. Oswald
1982—Milton Derber
1983—Jack Stieber
1984—Wayne L. Horvitz
1985—Everett Kassalow
1986—Lloyd Ulman
1987—Michael H. Moskow
1988—Phyllis A. Wallace
1989—Joyce D. Miller
1990—Robert B. McKersie
1991—James L. Stern
1992—Ernest J. Savoie
1993—George Strauss
1994—Lynn R. Williams
1995—Walter Gershenfeld
1996—Hoyt N. Wheeler
1997—Francine D. Blau
1998—F. Donal O'Brien
1999—Thomas A. Kochan
2000—Sheldon Friedman
2001—Magdalena Jacobsen
2002—John F. Burton, Jr.
2003—Paula B. Voos
2004—Marlene Heyser
2005—Stephen R. Sleigh
2006—David B. Lipsky
2007—Eileen B. Hoffman
2008—Anthony Oliver, Jr.
2009—Joel Cutcher-Gershenfeld
2010—Eileen Appelbaum
2011—Gordon Pavy
2012/13—David Lewin
2013/14—Lisa Lynch
2014/15—Martin Mulloy
2015/16—Bonnie Castrey
2016/17—Janice Bellace
2017/18—Harry Katz
2018/19—Kris Rondeau
2019/20—Dennis Dabney
2020/21—Adrienne Eaton
2021/22—Wilma Liebman

# LABOR AND EMPLOYMENT RELATIONS

# ASSOCIATION SERIES

# Proceedings of the LERA 2023 Meetings

**LERA@ASSA Meeting**
**January 6–8, 2023, New Orleans, LA**
**(in conjunction with ASSA/AEA)**

and

**LERA 75th Annual Meeting,**
**June 1–4, 2023, Detroit, MI**

**J. Ryan Lamare, Editor-in-Chief**

# CONTENTS

## VI.
## LERA ANNUAL REPORTS

# I. Presidential Address

## Making Work a Positive Part of the Human Experience—For Everyone

**PAUL CLARK**
*Pennsylvania State University*

In preparing this talk, I sought advice from a number of people about how to present an address of this kind. I repeatedly heard three suggestions—go big picture, make it personal, and keep it short. I intend to follow all three suggestions.

The focus of my remarks today is "work." I think that is a sufficiently "big picture."

Work is a fundamental part of the human experience for the vast majority of people in our world. Yet we have not made enough progress in making work a meaningful and fulfilling part of everyone's life.

As a society, we have made tremendous progress in just the past few decades in medical science, information technology, and space exploration. We deciphered the genome, we created the Internet, and we send spacecraft to Mars, Jupiter, and Saturn, but we have not made the same kind of progress in the role work plays in our lives.

LERA is a unique organization that brings together practitioners and academics who, in many different ways, play an important role in shaping the experience people have at work. Our membership includes arbitrators and mediators, labor relations and human managers, union representatives, attorneys, people employed by government agencies, and academics. Work, and the employment relationships related to work, is what brings us together. If fundamental change needs to occur in the area of work, LERA and its members are well placed to lead the effort.

An average American employed full-time will work for pay 90,000 to 100,000 hours in the course of a lifetime. The only other single activity many of us might spend more time on is sleep. So, work is a big part of our lives. The pay we receive for our work largely determines our standard of living, but work also has the potential to meet many of our important psychological and social needs.

Work can be, and I would argue *should* be, a positive experience in people's lives. But it isn't for everyone—at least according to that famous social scientist/comedian, George Carlin, who said about the role of work and employment in people's lives: "Oh, you hate your job? Why didn't you say so? There's a support group for that. It's called everybody, and they meet at the bar." Of course, that is an exaggeration. For many people, work is a positive experience. But in our world today, how many people *do* hate their jobs? Or, to put it a different way, how many people do not see their work as a positive experience that enriches their lives?

Now, I will segue to the "make it personal part."

I grew up in a steel town near Pittsburgh. The people in my community worked mostly at the local mill or at the plant that made railroad cars. I was not from a blue-collar family (my father was a professional Boy Scout for most of his life—that is actually a job). But every day, our news was filled with stories about unions, strikes, and layoffs. So, growing up, I was constantly reminded that work and employment played an important role in people's lives. My awareness was further heightened when, at age 18, I went to work in a glass factory.

My siblings and I were the first generation in our family to go to college, but we had to largely pay our way (professional Boy Scouts don't make much money). That necessitated getting a serious job that paid more than typical teenage summer jobs like mowing lawns or caddying at the local golf club. So even before I graduated from high school, I began picking up shifts on weekends in a factory that made glass bottles. And I would do so every summer, and during most breaks from college, for the next five years.

Working in the glass factory was not considered a particularly good job. It didn't pay as well as the steel mills and the coal mines. But I still made double what most other kids my age were making. The factory was hot, noisy, and smelled of sulfur. And since we worked with molten glass, it could be dangerous. I worked (variously) all three shifts. The work itself was physically taxing, but the hardest part was the monotony.

Those of us on the production line were assigned a place to stand and told what to do, and everybody was expected to do the job as instructed. And the line moved very fast. The supervisors watched us closely and would yell at us if we didn't keep up. When I later went to graduate school, I was able to put a name to this management approach: "Oh, so breaking down a task into its simplest parts and making sure that everyone does the job the way the boss tells you has a name? Taylorism? Who knew?"

And the fact that the supervisors watched our every move to keep us from loafing represented, I later learned, a "Theory X" view of workers. Of course, all of this would be news to them because they were just doing what the supervisors before them, and the supervisors before them, had done.

In addition to learning firsthand about unions (I was a member of the Glass Bottle Blowers Association—that's actually what the union was called). I also quickly learned that most workers in the factory really disliked their jobs, but they still dutifully showed up day after day to pay the bills. I remember wondering how many people beyond the glass plant had such a negative experience at work.

In 2012, a major study was conducted that examined work in 1952 and what it was like for workers in this country 60 years later. The report concluded that the world of work had changed between the 1950s and the 2010s, "but it was not a change which is making many of us happy." Overall, the report said work continues to be "the warp and weft of everyday life." What does warp and weft mean? Well, these are terms that come from the weaving of cloth, but the phrase translates to "unpleasant things that happen over and over again." In the 21st century, is work for many people still an unpleasant thing that happens over and over again? Unfortunately, there is lots of evidence that it is, both in the United States and around the world.

Here is a headline from 2022: "Job unhappiness is at a staggering all-time high, according to Gallup." Gallup polling showed that workers were experiencing high rates of disengagement and unhappiness. The report found that 60% of workers globally were emotionally detached at work, 19% were miserable, and only 21% reported feeling engaged.

Another indicator of how people feel about their jobs is the percentage of people who want out. A 2023 study found that 61% of US workers want to leave their current job.

So, to review, when asked how they feel about their work, large percentages of employees tell us they are disengaged, unhappy, emotionally detached, miserable, and want to quit.

You might argue, at this point, that employers are not required to make work engaging, satisfying, or meaningful. And, of course, that is true. But there is also evidence that it is very much in an employer's self-interest to do so.

Let's just look at employee engagement. Gallup has found that the profits of businesses with engaged workers are 23% higher than businesses with disengaged workers and that these workers have significantly lower absenteeism, turnover, and accidents. The Hay Group, a consulting firm, reports that "companies with a highly engaged workforce outperform firms with the most disengaged workforce by 54% in employee retention and by 200% in revenue growth."

Finally, to look at this from a national perspective, Gallup found that disengaged workers cost the United States $450 billion to $550 billion each year as a result of increased turnover and lower employee productivity.

So, the evidence is compelling. If work is made more engaging, employees have a much more positive experience at work, and companies benefit in significant ways. Employers similarly benefit from increases in worker voice and job satisfaction.

While some employers have made improving the quality of work for their employees a priority, many employers have chosen not to do so. In the future, employers may not have a choice in the matter. Why? Because there is evidence that the newest generation of workers will demand that employers make improving work a priority.

So, let's talk about Generation Z. Generation Z comprises people born between 1996 and 2010. The oldest members of this generation are in their twenties and make up the youngest part of today's workforce. Research indicates that this generation has higher expectations in terms of what they want out of work than previous generations. And this appears to hold true, regardless of whether they do white-collar or blue-collar work, or what sector they work in.

When I think back to my time in the glass factory, I remember that the workers there seemed to stoically accept their situation; they didn't have expectations that a job should provide anything more than a paycheck. Generation Z sees work in very different terms.

So, what does Generation Z want from work? Recognizing that no group of workers is monolithic, there are five things that are consistently mentioned in the research that are priorities for these workers—voice, purpose, flexibility, well-being, and social responsibility.

It should not surprise us that many in this new generation of workers want a greater voice at work. I am sure many folks here are familiar with the work of Kochan, Yang, Kimball, and Kelly on worker voice and the voice gap. Kochan and colleagues found that up to 60% of workers surveyed experienced a difference between their expected level of voice and their actual level of voice. The evidence suggests that many Generation Z members also feel they have insufficient voice in their workplaces, and they are demanding more.

One manifestation of this desire for greater voice is likely the growing number of organizing campaigns we see in sectors heavily populated by younger workers. Just in the past few years, we have seen organizing efforts at Starbucks, Amazon, Apple, Trader Joe's, REI, and Chipotle. Graduate students and part-time faculty are organizing all over the country, as are medical residents and interns.

Many of the Starbucks workers point to a greater voice in their work lives as one of the driving forces behind their organizing efforts. One barista said, "It doesn't seem fair that higher-level management has a greater say in coronavirus safety protocols than the people brewing coffee in person every day."

In a recent interview, some Starbucks workers said they like the company and the values it stands for, and they see a union as a way of getting its leaders to live up to those values. An 18-year-old barista in Ithaca, New York, was asked why she wanted a union. She said, "We want to be able to sit down with Starbucks, with the higher-up executives, and make a plan so that we, as employees, feel as valued as they say that we are."

A 2023 article in the *Harvard Business Review* describes Gen Z as a purpose-driven generation. It goes on to say that what differentiates them from other generations is a desire to know how their work contributes to the mission of the organization, as does the fact that their career and consumer decisions are based on the impact those decisions make in the world. This desire for meaningful work, for purpose, presents a significant challenge for employers.

And not surprisingly, the newest generation, having lived through the pandemic, wants flexibility at work. Surveys show that having options to work in the office, at home, or in some combination, is a high priority for this generation. They want flexible work arrangements to help deal with health and family challenges. And many want to do away with the traditional workday and workweek. Many employers are already moving in those directions.

Gen Z workers are also concerned with their own well-being. They are likelier to struggle with job stress, and because they are way ahead of other generations in destigmatizing and openly talking about mental health concerns, they are pressing employers to place a much greater emphasis on mental and emotional health care and work–life balance.

There is evidence, too, that Gen Z'ers believe in corporate social responsibility and want to work for companies that share their beliefs and values. They are strongly committed to diversity, equity, inclusion, and belonging. And while parts of society are pushing back on corporate initiatives in environmental, social, and governance, Gen Z workers want to work for companies that are committed to these principles.

In sum, Generation Z workers want a different kind of workplace than the one their parents experienced. And I don't think they are talking about a workplace where everyone does what they want, and no one is accountable. Not at all. I think they understand that accountability is an integral part of a positive work experience.

In any case, change in the way we do work is inevitable. Changing technology, and in particular, artificial intelligence, will make certain of this. As we reorganize work, I hope we make it more engaging, more meaningful, and more flexible for everyone. We already see changes that can have this kind of impact.

Some employers have adopted "high-road employment practices" that benefit employees and employers alike. Labor–management partnerships greatly increase voice and engagement in unionized workplaces. The adoption of the four-day workweek and the provision of childcare and paid family leave by some employers reduce the stress on workers with families. We need more employers to adopt such changes and go even farther if we are to meet the expectations of Generation Z.

I also wonder whether those of us engaged in the practice of teaching and research have been too patient in waiting for meaningful change regarding the role of work in people's lives. I think it is time to be much more aggressive in bringing about change in this area.

I began this talk by mentioning the range and scope of expertise represented among LERA's membership. How can we use that expertise to accelerate innovation in the ways we organize work in the United States and beyond? How can we make fundamental advances in this area that are on the scale we have seen in the medical sciences, information technology, and space exploration? How can we lead the way in making work a positive experience for all? It's a very big challenge, but it's also a great opportunity.

In closing, I want to say that I very much appreciate the opportunity to have served as LERA president this past year. I have been a LERA member, both at the national and chapter level, since 1978. And the association has been a very important part of my professional life. I would encourage folks to invest time in your chapters and the national organization. The returns on that investment will be worth the time and effort.

Finally, I would like to say that one of the things LERA presidents get to do during their term is look behind the curtain and see how our association functions on a day-to-day basis. Sometimes, looking behind the curtain can be a very scary thing, but in the case of LERA, it's very reassuring. As I think everyone here is aware, we have wonderful staff who really "get" what LERA is about and who are truly committed to helping the association be the best it can be. So please join me in thanking our LERA staff—Emily Smith, Bernadette Tiemann, and Michael Lillich.

# Discussant Comments for Session 12.5: Historical and Theoretical Perspectives

**ZACHARY SCHALLER**
*Colorado State University*

This session brought together scholars from a diverse set of disciplines and featured an equally diverse set of research methodologies, which is pleasantly typical of a LERA event. Together, their historical and theoretical perspectives (often bolstered by empirical evidence) covered a delightfully rich set of topics, most notably:

- the important role that unions play in society as prosocial and libertarian (yes, even libertarian) organizations
- the relevance of historical moments such as 1980 to the trajectory of economic outcomes and geospatial polarization in the United States
- the intensifying divide between economically anachronistic (mostly rural) regions and economically progressive urban centers (mostly coastal cities).

In his paper titled "The Relevance of Institutional Economics for Labor Economics Through Understanding American Economic History," Jerome Braun provides a review essay that situates Jonathan Levy's book, *Ages of American Capitalism: A History of the United States*, within the frameworks of institutional and neoclassical economics. History of thought scholars will find a lot to like in Braun's work. He effectively incorporates the ideas of the old institutionalists, such as Thorstein Veblen and John R. Commons, to offer a contemporary critique of neoclassical economics. He boldly criticizes neoclassical economists for prioritizing models over empirical evidence and market predictions over behavioral observations—criticisms that are often raised in present-day discourse but are not always recognized as key features of the old institutionalists. It is funny how everything comes back into style with enough time.

The influence of this paper might expand if it explicitly connects the old institutionalists with the new institutionalists, particularly mentioning the influential work of Douglas North. By doing so, Braun can bridge the familiarity gap for recently minted economists who may not be well acquainted with Veblen and Commons but are likely to be familiar with North's contributions. Additionally, an effective approach to critiquing neoclassical theory might involve using the language and framework of neoclassical modeling. Most, if not all, economists recognize the fundamentals of modeling, and many are receptive to the friendly reminder to make realistic assumptions when they are critical (meaning pivotal to the prediction) and to match models with their appropriate context. I have in mind here the work of Dani Rodrik in *Economics Rules*.

Braun's essay stands out for its strong stance against excessive corporate power and its critique of modern capitalism's disconnection from community values. However, this discussion raises important questions about the extent to which flawed economic ideas can be held responsible for a broken system and the path of economic history. Are these ideas directly accountable for the current state of affairs, or have they simply contributed to the increasing dominance of corporate power? Exploring this sensitive question can sharpen our focus on the rules of the game that need to change and the broader consequences of economic theories.

For the second presentation in the session, Mark Reiff delivered a thought-provoking talk about his book, *Unionization and the Argument from Liberty*. Reiff skillfully employs a libertarian thought experiment to explore the role of unions in a free society and the permissibility of union shops within that framework. His work offers a broad and ambitious perspective that encourages scholars to move beyond incremental policy changes and consider how the fundamental rules of the game can be transformed to benefit workers and society as a whole.

A key focus of labor relations revolves around right-to-work laws and the constitutional validity of fair-share fees. Interestingly, Michigan recently repealed its right-to-work law, making the topic particularly relevant. So-called right-to-work laws are often advocated for through a libertarian argument, asserting that individuals should not be compelled to join a union or pay fair-share fees due to their right of association, including the right to choose not to associate with a union. Reiff wastes no time in addressing this argument head-on. In a creative twist, he flips the script to show that universal union coverage can and would emerge within a libertarian utopia. This challenges the prevailing narrative that libertarianism inherently opposes unions and union shop arrangements.

Reiff builds his case by drawing on the ideas of Robert Nozick, who argues that in a libertarian utopia, "dominant protective associations" would naturally emerge. Nozick contends that these associations, essentially akin to a minimal state, would encompass universal membership as a consequence of libertarian ideals. Reiff skillfully guides readers through an extensive philosophical and economic discussion, addressing potential objections along the way. However, the most pivotal and debatable aspect arises when considering the treatment of "independents" who choose not to join the union. Nozick's perspective on this matter contends that independents engage in "risky behavior," which violates the rights of others. Reiff adapts this reasoning to argue that employees who opt out of union membership impose similar risky behavior that could violate the rights of others within a specific workplace and undermine unions in the utopia.

The argument is novel and thought provoking. However, it is important to note a significant distinction between a union organization and a dominant protective association, which may limit the legitimacy of extending the argument. One way of viewing mandatory membership in a dominant protective association is that it is equivalent to citizens putting on the bonds of civil society under a monopolist on violence, to borrow the language of Locke and Weber. Yet asking citizens to divest themselves of their natural liberty becomes less convincing when applied to an organization that is not the minimal state. If we accept the necessity of compulsory union membership for employees in a specific company, does it follow that membership in other organizations could also be mandated, provided we define the appropriate population scope? In essence, what makes unions uniquely justifiable for compelled association? One might be concerned that other organizations could exploit this logic, manipulate it for their own purposes, and expand compulsory memberships.

Despite my reservations over certain elements of the argument, I found many parts of the book to be tremendous. One notable example is Reiff's adept incorporation of Ronald Coase's insights into the political economy of labor relations. Coase's observation that "the distinguishing mark of the firm is the suppression of the price mechanism," when combined with Elizabeth Anderson's assessment that "we are told that our choice is between free markets and state control, when most adults live their working lives under a third thing entirely: private government" presents a compelling case for the libertarian necessity of unions.

Additionally, Reiff eloquently explores the inherent dignity of labor and the human aspect of the individuals providing those services. He presents a compelling argument that conventional supply and demand models of the labor market unjustly overlook these distinctive qualities, reducing workers to mere mediators of a production commodity. Compassionately, he asserts that "people, unlike products, have roots." Workers cannot be treated as commodities subject to market competition, which is why unions do not engage in anti-competitive behavior.

Overall, Reiff's book is a fun read and serves as an inspiring call to approach the political economy of labor relations with fresh and expansive thinking.

The last two papers in the session shifted the focus and methodology toward empirical insights. Scott Abrahams and Frank Levy contributed "Run-Up to Populism: A 39-Year US Perspective." The paper cleanly and convincingly connects important US trends in economic production, educational attainment, spatial inequality, and cultural attitudes. The authors effectively illustrate the interconnectedness of these typically fragmented areas in the literature, specifically highlighting how the decline in manufacturing since 1980 and the subsequent increase in college attainment have contributed to heightened spatial inequality and political polarization. A key finding is that regions that were heavily reliant on manufacturing in 1980 remained so in 2019, albeit at a reduced scale due to changes in the national economy. Similarly, areas with a high proportion of college-educated individuals in 1980 continued to attract and retain college graduates in 2019, resulting in intensified educational disparities. This path dependence has fueled significant spatial inequality, particularly in the rural–urban divide, and has generated cultural divisions as rural workers feel increasingly disconnected from the college-educated "coastal elite." Populism has been a natural outgrowth of these fissures.

The authors deserve credit for avoiding the oversimplification of attributing these trends to a single cause, which would be reductionist, and for refraining from making definitive causal claims, which would be overly ambitious. Instead, they carefully establish associations between the trends and highlight the intuitive connections among them. Their rigorous approach is commendable, and their contribution is praiseworthy.

There are two implications of this study that I will highlight. The first is that cultural attitudes surrounding higher education in rural places should be very concerning. Combined with the path dependence that has left these places largely devoid of new investment and retooling, the United States is in grave danger of creating poverty traps in its own backyard. The mental association that people have formed (whether real or imagined) between college attainment and left social values has created a peculiar distaste for higher education in rural communities. It is understandable from an evolution of identity and culture perspective, but it is occurring at great cost to the nation and its rural communities—not to mention the potential for great personal cost to many rural youth. A quiet deskilling and divestment in America's interior is a crisis in the making. Populism and certain expressions of right-wing extremism are unfortunately only the tip of the iceberg.

The second implication follows from the first. Organizations of all kinds, but most notably academic institutions, are increasingly losing diversity of thought because of spatial inequality in educational attainment. Diversity, equity, inclusion, and belonging initiatives often focus on race, sexual orientation, and gender (for many excellent reasons), but they fail at times to account for socioeconomic status and regional diversity, two important factors determining diversity of thought. It is worth thinking about the costs of losing poor rural experiences from policy forums and academic discourse.

Finally, Jack Fiorito presented "Why Do People *Really* Support Unions?," a paper coauthored with Andrew Keyes and Zachary Russell. This work summarizes a series of recent studies performed by the authors and other collaborators focused on prosocial motivations for union support. It has long been known that workers care about workplace instrumentality—a union's ability to deliver improved wages, benefits, and voice. And there is empirical evidence in that vein, such as Kochan (1979). Unfortunately, there is substantially less work on prosocial motivations for union support. The authors channel the spirit of Allan Flanders (1970) by maintaining "that unions must be a 'sword of justice' at work and in society to generate loyalties and induce sacrifices among members that are foundations of unions strength and vitality." To what extent workers are motivated to support, join, and find satisfaction in unions is an open empirical question that the authors answer convincingly.

Through a series of survey studies, most featuring an experimental vignette method, the authors present compelling evidence that individuals highly value the prosocial impact of unions. Their preliminary statistical analysis indicates a strong positive association between experimental cues emphasizing prosocial factors and various support measures, including general union attitude, prounion voting intent, union approval, and even union satisfaction among workers with union experience. In fact, the estimated effects of these cues are often comparable to or larger than those related to workplace instrumentality. These findings suggest that if unions can enhance workers' perceptions of prosocial benefits, it could lead to meaningful increases in approval of unions and intent to vote for union representation (I refrain from sharing exact figures out of respect for the preliminary nature of the results. For current estimates, see the authors' working papers).

A potential challenge in interpreting and applying these findings relates to the subjective nature of the concepts "prosocial" and "social justice." Different individuals may hold distinct views on how these concepts should be expressed in practice. For instance, workers in Waco, Texas, may advocate for social justice policies that workers in Atlanta, Georgia, might perceive as conflicting with their own understanding of social justice. Renewed debates about abortion access within healthcare or affirmative action within higher education come to mind. Unions dedicated to advancing social justice, and enjoying increased member support because of it, should prioritize understanding their members' perspectives on social justice. This understanding is essential to prevent inadvertent misinterpretations or misalignments with members' expectations.

Two additional implications arise from the research. First, unions continue to serve as a crucial haven for civic engagement in the United States. At a time when traditional civic organizations are experiencing declining participation and people feel further disconnected from community affairs, unions stand out as a rare domain where individuals can actively participate in collective endeavors and champion the welfare of others. Fiorito, Keyes, and Russell's findings highlight the enduring value that people place on unions as a platform for engaging in prosocial initiatives. As Robert Putnam discusses in *Bowling Alone*, unions play a crucial role in fostering community engagement. They often extend their activities beyond the workplace to support local communities through initiatives such as organizing charity events, participating in community service projects, and advocating for social causes that benefit not only their members but also the broader society. By actively involving their members in these endeavors, unions strengthen the bonds among workers, their communities, and societal issues. Despite reduced prevalence, there remains strong potential for unions to perform this important function in society.

Last, prosocial motivations may be a larger part of the COVID union revival than commonly thought. Yes, the spike in union activity has a lot to do with worker voice and concerns over workplace instrumentality, but this paper forces one to consider that it also has a lot to do with manifesting change in economic and social justice. Where COVID kept us apart physically, it may have brought us together as a community of fate.

## References

Anderson, Elizabeth. 2017. "When the Market Was "Left." In *Private Government: How Employers Rule Our Lives (and Why We Don't Talk about It*, pp. 1–36. Princeton, NJ: Princeton University Press.

Flanders, Allan D. 1970. *Management and Unions: The Theory and Reform of Industrial Relations.* London, UK: Faber.

Kochan, Thomas A. 1979. "How American Workers View Labor Unions." *Monthly Labor Review* 102, no. 4: 23–31.

Putnam, Robert D. 2000. *Bowling Alone: The Collapse and Revival of American Community.* New York, NY: Simon & Schuster.

Rodrik, Dani. 2015. *Economics Rules: The Rights and Wrongs of the Dismal Science.* New York, NY: Norton.

# III. LERA Lifetime Achievement Awards

**GREG J. BAMBER**
*Monash Data Futures Institute*

Thanks, President Paul Clark. It is a great honor to receive this award from the Labor and Employment Relations Association, the leading professional association in our field, presented by you. I was lucky enough to become friends with you, and with many other lovely people through LERA. This highlights valuable aspects of LERA: it is a wonderful forum for learning, the exchange of ideas, and getting to know like-minded people! You contribute your astute leadership skills and fine values to Penn State University, as well as to LERA. Many congratulations to Janice Bellace and Steve Sleigh for their Lifetime Achievement Awards.

I am especially grateful and pleasantly surprised to be offered this award because, in the history of these awards, LERA has conferred such an award on only two other people who were not based in the United States.

Some people say that our field of industrial relations is dying. We must not let it die but must continue to breathe new life into this vital field. We were reminded of its importance during the COVID-19 pandemic when we saw new types of workplace disputes that deserved novel forms of conflict resolution (Teicher, Van Gramberg, and Bamber 2023).

Industrial relations should not be dying! It's more important than ever. On the one hand, billionaires send people joy-riding into space, while those who generate their wealth have precarious jobs that all too often are *not* decent forms of good work. There are too many people in jobs that are unsafe, unhealthy, managed by algorithms, and under surveillance by artificial intelligence.

Too many people are subject to wage theft, are paid starvation wages, and may not be allowed any voice about their work, including through unions—which may have to confront ruthless union-busting employers and consultants. Many workers are not allowed enough time for breaks, are subject to discrimination or harassment, and they may also be homeless and face massive inequalities in terms of pay and conditions.

An indication that industrial relations is not dying is that there are so many interesting sessions at LERA meetings that we are often spoilt for choice! That's a problem of success. The students who are fortunate to be here will carry our field forward and develop it further to help confront the many areas of society that need to be improved, as exemplified by many of the vital topics that we're discussing in LERA. Let's continue to try to improve our employing organizations, institutions, theory, practice, regulation, and thence our society. There are still many challenges for us to confront, and much work for us to do.

Let's try to apply our learning from LERA and elsewhere to the many aspects of society that need to be improved in the United States and around the world. There's so much that we can learn from and contribute to other countries.

When I was a student, I decided to major in industrial relations. It was then high on the political agenda in the United Kingdom and some other countries. Unions were growing, and there was much discussion of workers' rights and industrial democracy. The academics I was lucky enough to be taught by were passionate about theory, practice, and policy. Inspirational scholars led the research areas where I was a student. They included George S. Bain and Robert H. Fryer at Manchester University; Ben R. Roberts, Keith E. Thurley, and David Winchester at the London School of Economics; and Thomas L. Johnston and Alexander I. I. Wedderburn at Heriot Watt University, Edinburgh.

Fortune also provided me with kind mentor-friends, including Peter D. Carr, John Goodman, Roderick Martin, and Gill Palmer. Several of those mentioned were members of LERA's predecessor; this encouraged me to start attending its meetings, which I continued to enjoy after moving to work in Australia.

Not only have I enjoyed participating in LERA, but also in its United Kingdom and Australian counterparts, and what is now called the International LERA, which holds interesting congresses around the world. Janice Bellace, Harry Katz, Tom Kochan, Russell Lansbury, and several other fine LERA members have skillfully served as its presidents.

If others see me as having any achievements, they might be referring to my co-publications—for example, on international and comparative employment relations (e.g., Bamber, Cooke, Doellgast, and Wright 2021), new technologies in workplaces (e.g., Budhwar et al. 2023), unions (e.g., Clark et al. 2021), and airlines' and hospitals' management of people (e.g., Bamber, Gittell, Kochan, and von Nordenflytch 2009; Stanton, Bartram, and Bamber 2022).

To misquote Isaac Newton, any such achievements are based on working with teams and standing on the shoulders of giants, many of whom were LERA members and who helped me in various ways. Apart from those already mentioned, there are Alex, Anil, Bob, Bonnie, Darlene, David, Eileen, Joel, Kathy, and .... Please forgive me; there are too many friends and colleagues to mention who have been integral to my modest achievements.

Thank you again, LERA, for this honor! Emily, Bernadette, and Mike: LERA seems to run like a well-oiled machine, largely thanks to you. For indulging my lifelong curiosity about industrial relations and related fields, I also thank Dale, Alex, Kate, Leuca, Coco, and the rest of my family and friends. I also acknowledge my late parents, Betty and Doug—who, when I was a child, were active in the UK's National Union of Teachers.

## REFERENCES

Bamber, Greg J., Fang Lee Cooke, Virginia Doellgast, and Chris F. Wright, eds. 2021. *International and Comparative Employment Relations: Global Crises and Institutional Responses*, 7th ed. London, UK: SAGE Publishing.

Bamber, Greg J., Jody Hoffer Gittell, Thomas A. Kochan, and Andrew V. von Nordenflytch. 2009. *Up in the Air: How Airlines Can Improve Performance by Engaging Their Employees,* Ithaca, NY: Cornell University Press.

Clark, Paul F., Greg J. Bamber, Paul V. Whitehead, Lois S. Gray, Sandra Cockfield, and Kay Gilbert. 2021. "Does Modernizing Union Administrative Practices Promote or Hinder Union Revitalization? A Comparative Study of US, UK and Australian Unions." *British Journal of Industrial Relations* 59, no. 2: 370-397.

Budhwar, Pawan, Soumyadeb Chowdhury, Geoffrey Wood, Herman Aguinis, Greg J. Bamber, Jose R. Beltran, Paul Boselie, Fang Lee Cooke, Stephanie Decker, Angelo DeNisi, Prasanta Kumar Dey, David Guest, Andrew J. Knoblich, Ashish Malik, Jaap Paauwe, Savvas Papagiannidis, Charmi Patel, Vijay Pereira, Shuang Ren, Steven Rogelberg, Mark N.K. Saunders, Rosalie L. Tung, and Arup Varma. 2023. "Human Resource Management in the Age of Generative Artificial Intelligence: Perspectives and Research Directions on ChatGPT." *Human Resource Management Journal* 33, no. 3: 606–659.

Stanton, Pauline, Timothy Bartram, and Greg J. Bamber. 2022. "The Role of Nurses' Unions in Workplace Innovation in Australian and Canadian Hospitals: Analysing Union Strategies." *Industrial Relations Journal* 53, no. 5: 484–500.

Teicher, Julian, Bernadine Van Gramberg, and Greg J. Bamber. 2023. "Understanding Workplace Conflict and Its Management in the Context of COVID-19." In *The Evolution of Workplace Dispute Resolution: International Perspectives*, edited by Ariel C. Avgar, Deborah Hann, Ryan Lamare, and David Nash. Champaign, IL: Labor and Employment Relations Association.

# III. LERA Lifetime Achievement Awards

**JANICE BELLACE**
*Wharton School, University of Pennsylvania*

I am honored to receive the LERA Lifetime Achievement Award, but I must confess that I was most unexpected. Me? The person who felt a little like the "odd man out" when I first joined LERA?

I had done my undergraduate and law degrees in the United States and greatly enjoyed labor law, especially with Bob Gorman as my teacher. I assumed I would first go to the National Labor Relations Board and then go into practice. But I thought that before doing that it would be a good idea to study industrial relations. I applied and was awarded a fellowship and went to the London School of Economics. I had no inkling that this would change my entire view of labor law. At the London School, I learned about US industrial relations from British professors, such as Ben Roberts, and instead of reading works of the major US scholars, I read the works of British scholars. It was happenstance that Britain had just joined the European Community and my labor law professor, Lord Wedderburn, was keen to discuss how the labor laws in European countries were so different. Quoting the famous Sir Otto Kahn-Freund, he stressed that it was impossible to understand another country's labor laws unless one had some understanding of the industrial relations system, the history, and the political structure in that country.

On returning to the United States, I did not immediately go into practice. My husband was doing an MBA at Wharton, and Herb Northrup, head of the Industrial Relations Unit there, offered me a research position. Once again, this led to a major change in my career path. I realized that I loved research—and Wharton gave me the opportunity to do research in international and comparative labor law and labor relations. Soon after I started at Wharton, Bernard Samoff in the Management Department told me that I should join LERA (then IRRA). I began to attend the monthly meetings of the Philadelphia chapter, where I met wonderful people like Walter and Gladys Gershenfeld. Bernie also told me that I should go to the annual meetings. I said that I knew no one and was not on the program. In his direct way, Bernie said "Go. Meet people. And submit a paper next year." So I went to Atlanta, did meet people, and did submit a paper the next year. Thus began my 40-plus-year involvement with the national LERA.

Why did I feel the odd man out? Because in going to LERA, I realized that nearly everyone else, with doctorates in economics or industrial relations, did a different type of research than I did as a labor law academic. I noticed that there was a Cornell crowd, a University of Wisconsin crowd, and later an MIT group. There was no Penn group. But people were welcoming, and over time I got to know well many people who have become friends. Another oddity for me was that in the early 1980s, there were almost no sessions that were not about US industrial relations. When international sessions started, they were often relegated to the 8 a.m. Sunday slot because low attendance was expected. Once again, over time, things changed—and today, a good number of sessions have an international dimension.

My own research over my career has moved. Otto Kahn-Freund once wrote that the purpose of labor law is to support, restrain, and regulate the power of management, with the aim of improving the terms and conditions of employment. At the start of my academic career, I focused on US labor law and some issues relating to gender equality—in particular, equal pay. By the 1990s, I was writing comparative pieces, for instance, comparing the bargaining structures in the United States, Germany, and Japan. )Needless to say, I found the sectoral approach of the latter two countries preferable.) In the mid-1990s, I was appointed to the International Labour Organization's Committee of Experts, a group of 20 academics and jurists from around the world who examine member states' compliance with ratified conventions. This affected my thinking about labor law. I now saw it as a way of establishing "rules of the game"—that is, standards to be observed with the goal of achieving social justice.

I was on the ILO Committee of Experts at the time the 1998 Declaration of Fundamental Principles and Rights at Work was adopted. In one of my articles soon after I pointed out that, in effect, the ILO declaration, which did not use the term "human rights," set out four foundational human rights that had particular applicability at work. Over the following decade, those rights would be included in the UN compact, in hundreds of company codes of conduct, in labor clauses in free trade agreements, and finally in the UN Human Rights Council's Guiding Principles on Business and Human Rights. My writings on this culminated in a book, *Labour, Business and Human Rights Law*, in 2019. I still recall the British publisher being perplexed by the title, asking why "business" was included. I said that if workers have rights, someone has a duty to respect those rights.

More recently, I have written about labor clauses in trade agreements because concern has increased about labor exploitation and human rights in global supply chains. I have also returned to some basic themes but with a different lens. My most recent book (co-authored) is titled *The Right to Strike in International Law*. From a pragmatic angle of analysis, we all know that without the right to strike, workers can only engage in collective begging. From a human rights angle, we also know that workers' freedom of association, which is critical to democracy, means little without the right to strike.

I am honored to receive this along with Steve Sleigh, whom I met years ago, and Greg Bamber, whom I know from my British connection. I could name so many people I have come to know, admire, quote, and enjoy through LERA. I refrain because it would begin to sound like the Academy Awards. So rather than leave anyone out, let me simply say that what makes LERA great is its purpose and the people. LERA is the only association that I know that brings together academics and practitioners, and from different fields of study and practice. That's difficult to maintain, but LERA has done so because of the people—people willing to listen, to query, to challenge and, most importantly, willing to respect and engage with others. I have gained so much from LERA, and I hope that younger people will likewise see its enduring value.

# III. LERA Lifetime Achievement Awards

**STEVEN R. SLEIGH**
*Sleigh Strategy, LLC*

Thank you to LERA president Paul Clark and to incoming president Bill Spriggs, who I hope is feeling okay after a health setback. I am truly honored and humbled by this Lifetime Achievement Award, especially so to share the stage with Janice Bellace and Greg Bamber, two scholars I greatly respect and admire. I will come back to paying my respects to LERA but wanted to start out from the top by thanking Emily Smith and Bernadette Tiemann for the fantastic job they do. The past few years have tested all of us, but LERA pivoted quickly and effectively to do our meeting virtually and now returning in person. That is due in large part to the great work of Emily and Bernadette.

I describe my career in labor relations as chasing a thread. I'm on my sixth career over the past 50 years—and while not linear, there is a pattern. I first was exposed to the labor movement in high school while picking shade tobacco in Hadley, Massachusetts, during the summer. As the president of Youth for ATA (a Spanish acronym for a farmworkers 'group), I got to spend time at the Labor Center at the University of Massachusetts, Amherst, and to learn from Harvey Friedman, the long-time director there.

I met César Chávez during the tobacco worker organizing campaign in 1974—a real thrill to meet and interact with a living legend. After high school, I moved to Chicago and became a machinist, winning the Machinist Apprentice of the Year Award in 1978. Learning how to operate machinery such as a milling machine, surface grinders, and lathes was all new to me, and I'll confess really hard! Mastering the cutting of threads on a lathe—what's called chasing a thread—took me several years, but once I got it I had a great sense of accomplishment. My 11 years working as a machinist, the last seven of which were as a printing press erector setting up newspaper presses in seven countries and over 50 newspapers, was a great experience in my 20's.

It was during my Chicago years in the 1970s and 80s that I became active in the Machinists Union, participating in William Winpinsinger's Citizens Labor Energy Coalition. My tool box in 1979 was emblazoned with a big CLEC sticker that read "Labor Is Solar Powered"—maybe just a little ahead of time! I also had the great fortune to work with and learn from Ed Sadlowski, the head of the Steel Workers Union in Chicago, and a terrific sociologist, Bill Kornblum, whose book, *Blue Collar Community*, is one of the great ethnographies of a working-class community. After getting a scholarship from the IAM, I went back to study labor relations at UMass and then to Harvard's Kennedy School of Government. My advisor, John Dunlop, had written an arbitration ruling granting the IAM jurisdiction over printing-press installations within the AFL-CIO, while the Carpenters Union basically was granted jurisdiction over all other machinery installation. So, in a way, Professor Dunlop had been responsible for my opportunity as a printing-press erector.

At the Kennedy School, I took a labor economics class with James Medoff who, with Richard Freeman, had published *What Do Unions Do?* only a few years before. Not only was it a great class, but more importantly, I met Ann Greiner in that class; David Weil was our TA, and he helped Ann and I see that we belonged together! And he was right—Ann and I will celebrate our 32nd anniversary this fall.

After the Kennedy School, I had a brief interlude working in public policy for Governor Mario Cuomo but really was more interested in completing my academic arc and shortly started my Ph.D. studies at CUNY with my friend and mentor Bill Kornblum. It was while in New York in the late 1980s and early 1990s that I had the opportunity to work with Victor Gotbaum of AFSCME as the deputy director of the Center for Labor Management Policy Studies; Jack Sheinkman, president of the Amalgamated Clothing Workers, and Morty Bahr President of the Communications Workers of America.

I was enjoying teaching at Queens College when Ron Carey recruited me to become the director of research at the Teamsters Union in Washington, D.C., in 1992. We have lived there since then. I went back to my home union, the IAM, to create a new department, Strategic Resources, under George Kourpias and then Tom Buffenbarger—nearly completing my thread with labor leaders. At the IAM, I developed a unique approach to working with investors who were increasingly buying IAM-represented companies, leading me to an altogether different bend in my career arc: working as a partner in a private equity company buying union companies. Over the years, I have worked with some exceptional investors, including Ron Burkle, Michael Psaros, Adam Blumenthal, and Tim Fazio. Among them, they control companies that employ over 200,000 people and 100,000 union members. I pivoted again to head the IAM's pension and benefits programs, five years that struck terror in me every month that 100,000 checks were deposited in retirees' accounts. While there is a lot of sound and fury around the topic of environmental, social, and governance issues these days, the real job of running a pension plan is filled with a lot of block and tackling to make sure promised benefits are delivered.

Recently I am consulting with private investors who partner with unions in turning companies around. I had the opportunity to serve on the board of directors of the Amalgamated Bank, representing the Yucaipa Funds with a great partner and leader, Lynn Fox, the president of Workers United, the legacy to the Amalgamated Clothing Workers who founded the bank 100 years ago, and the chair of the bank's board. Lynn's union is running one of the most important union campaigns in recent memory—to organize Starbucks. The incredible energy and activism of the Starbucks workforce gives me hope that my long journey in labor is not focused on the rearview mirror but instead for what lies ahead. Part of that forward look is the work I do as co-chair of the Heartland Capital Strategy network. Together with Georgetown University, we created a fellowship program for college students who are interested in the intersection of labor and capital. Closer to home, my daughter Claire is beginning a master's program at Cornell ILR in August and my son James is busy planting gardens on office rooftops, making the world a better place.

I also want to acknowledge the great research partnerships I have participated in over the years with Tom Kochan, and especially Joel Cutcher-Gershenfeld. Research is what makes LERA great, as do all the participants from labor, management, and neutrals. Again, I deeply appreciate the Lifetime Achievement Award and all the work of LERA.

# IV. Tax Credits, Labor Supply, and Intergenerational Transfers

## The Effects of the Child Tax Credit on Labor Supply

**KYE LIPPOLD**
*Office of Tax Analysis, US Treasury*

### Abstract

The Child Tax Credit (CTC) is a major earnings subsidy in the US tax system, but it has received less research attention than the Earned Income Tax Credit (EITC). I identify the effects of the CTC by exploiting the fact that parents lose eligibility for the credit when a child turns 17. Focusing on the credit's effects among lower-income households in the Survey of Income and Program Participation, I find that loss of the credit leads to an 8.4 percentage point reduction in the probability that a child's parents are employed. This substantial response reflects the temporary nature of the lost subsidy; my results are consistent with a steady-state elasticity of 0.43, in line with previous EITC studies.

### Introduction

The Child Tax Credit (CTC) has been a consistently favored US tax policy for over 20 years, with all presidents since Bill Clinton adopting expansions to the credit. However, relatively little research has examined the effects of the CTC, despite its similarity as a wage subsidy to the Earned Income Tax Credit (EITC), a tax policy that has attracted consider stable interest for its effects on labor supply, health, and educational outcomes (Dahl and Lochner 2012; Hotz and Scholz 2003; Hoynes, Miller, and Simon 2015; Nichols and Rothstein 2016). In this paper, I examine the labor supply effects of the CTC as an initial step toward understanding how this policy has affected behavior.

Because the EITC and CTC have similar policy designs, my results for the CTC provide further evidence of the effectiveness of tax policies structured to encourage work. Policymakers continue to express interest in expanding credits like the EITC that subsidize earnings, given their positive effects on a variety of dimensions, but most studies of the effects of the EITC were performed using data from the 1980s and 1990s (Eissa and Hoynes 2004; Eissa and Liebman 1996; Grogger 2003; Hoynes and Patel 2018; Meyer and Rosenbaum 2001). There have been considerable changes in the economic and policy environment since that period, including welfare reform and the Great Recession, and some evidence suggests that labor supply responses have declined for more recent tax reforms (Heim 2007; Kumar and Liang 2016). Furthermore, recent research has questioned the previous literature on the labor supply effects of the EITC (Kleven 2021). Thus, it is valuable for policymakers to know whether labor supply responses to expansions of refundable tax credits have remained as strong as has been found in the past. Since the CTC subsidizes work similar to the EITC, it can provide evidence of these responses using new econometric techniques.

Another area of recent policy interest has been the projected effects of making the CTC fully refundable, as enacted for the single year of 2021 by the American Rescue Plan. Different researchers have found different conclusions regarding the labor supply effects of this policy, depending on assumptions (Acs and Werner 2021; Corinth et al. 2021; Goldin, Maag, and Michelmore 2022). This study provides empirical results about labor supply responses that are directly relevant to these questions. I focus on the "extensive margin elasticity" (the responsiveness of parents' decisions whether to work or not, rather than the "intensive margin" of varying their hours of work), as this has been the focus of previous EITC studies.

To identify the labor supply effects of the CTC, I exploit the fact that credit eligibility is based on the number of children under the age of 17 in a tax unit as of the end of the tax year. For example, a family with a child born on January 1, 1994, will receive a wage subsidy through the CTC for income in the entire year of 2010, while a family with a child born one day earlier on December 31, 1993, will receive no wage subsidy for that child in 2010. Other tax and transfer policies related to children typically change at ages 16, 18, or 19, but not at age 17. If other characteristics of parents evolve smoothly around the age 17 cutoff, comparing the employment rates of parents with children just above and below age 17 through a regression discontinuity (RD) design identifies the effect of the CTC on the labor supply.

In practice, the characteristics of children born in January and December differ due to seasonal factors, decisions related to timing of births, total years of exposure to the CTC policy, and incentives related to school starting ages (Buckles and Hungerman 2013; LaLumia, Sallee, and Turner 2015; Schulkind and Shapiro 2014). However, as long as these differences are consistent over time, it is possible to factor them out using a difference-in-regression discontinuities (DiRD) approach (Grembi, Nannicini, and Troiano 2016). This method combines a traditional RD and a difference-in-differences (DiD) model, using changes in treatment over time between the treated and control groups (in this case, children above and below the December birth month cutoff) to identify causal effects.

I estimate a DiRD design using data from the Survey of Income and Program Participation (SIPP), focusing on households from 2001 through 2016 with previous-year incomes below $20,000 (the range most subsidized by the CTC). My main results suggest an 8.4 percentage point fall in employment among low-income households with children over the age 17 cutoff, giving an extensive margin elasticity of 1.04. This value is an upper bound due to intertemporal substitution effects, but a rough calibration exercise suggests that my results are consistent with a steady-state extensive margin elasticity of 0.43. Thus, tax credits promoting employment for low-income workers continue to have substantial labor supply effects.

## Background

### *Design of the CTC*

The CTC arose out of tax reform in the 1980s, as members of Congress recognized that most other developed nations provided some form of child benefit to families (National Commission on Children 1991). Accordingly, several proposals were floated in the early 1990s to provide new tax deductions for families with children, and the CTC became law as part of the Taxpayer Relief Act of 1997 (Crandall-Hollick 2018). The original credit was effectively nonrefundable (i.e., unavailable to families that did not owe taxes), as it could only be claimed up to the amount of tax liability.[1]

A notable feature of the credit was the limitation of the tax break for families with children under 17. Negotiators in the House, Senate, and Clinton administration had proposed various age cutoffs for the credit, ranging from age 13 to age 18, but the final proposal adopted age 17 as a compromise (Esenwein and Taylor 1997). Importantly, this age 17 cutoff is unlike the other age cutoffs for child benefits in the tax code; parents can claim the dependent exemption and the EITC until their children turn 19 (or 24 when the child is a full-time student). This institutional detail is key to my identification of the credit's effects.

Four years after its enactment, the CTC expanded substantially as part of the Bush tax cuts of the early 2000s. The credit's refundable component, officially called the Additional Child Tax Credit (but referred to as the refundable portion of the CTC in this paper), was adjusted so that taxpayers owing taxes less than their maximum CTC could claim the remaining amount up to 10% of earnings above $10,000. One rationale for this initial $10,000 threshold was to counteract high marginal tax rates from the EITC at low levels of income; the policy intention was to have the CTC phase-in as the EITC phased out, lowering overall work disincentives (Sawhill and Thomas 2001). Further expansions in this vein followed; the subsidy amount was increased to 15% in 2004, and the threshold amount was lowered to $3,000 in 2009 (Crandall-Hollick 2018). Thus, the CTC has subsidized lower incomes in a manner similar to the EITC since 2001.

To illustrate how the CTC affects a family's tax liability during the period of my study, Figure 1 compares the 2016 amount of the CTC to the EITC for a single parent with two children, computed using the Taxsim model (Feenberg and Coutts 1993).[2] The two credits each subsidize earned income at a fixed rate (40% for a two-child

EITC, and 15% for the CTC) up to a maximum level, and they remain constant with income for a plateau period then are phased out (at a 21.06% and 5% rate, respectively) with increasing income. The CTC is smaller in value than the EITC but extends to higher income levels, meaning it provides its subsidy to most families in the population. The CTC also subsidizes earnings over a threshold ($3,000 in 2016) rather than subsidizing from the first dollar as for the EITC. The schedule of credits for other family arrangements is similar, as discussed by Lippold (2019).

**Figure 1. Value of CTC and EITC for a single parent with two children, tax laws for various years**

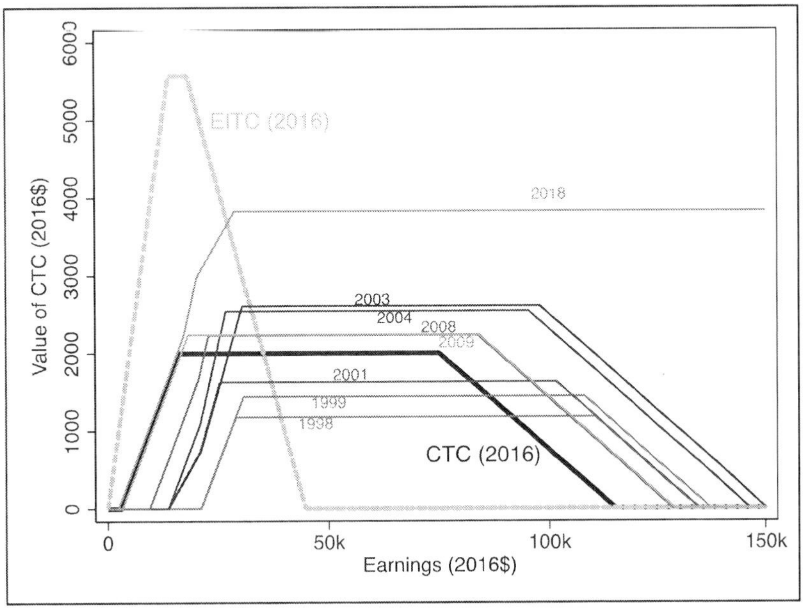


Source: Author's calculations using Taxsim. Calculations assume no unearned income or itemized deductions. Thin lines show CTC values for years where parameters were changed by law (before 2021).

Figure 1 also shows the changes to the CTC's design over time, with all values adjusted to 2016 dollars. The CTC has maintained a similar structure, with the value of the credit phasing in and out at similar rates; this held true even prior to the enactment of the refundable CTC because the credit offsets taxes in the lowest rate bracket. Also, the credit has fluctuated in value over time; its real value was lowest at enactment in 1998, rose in the mid-2000s due to the Bush tax cuts, and fell from 2003 to 2016 due to inflation. The credit's benefits shifted from wealthier families to lower-income families from 1998 to 2016 due to both (1) legislated reductions in the point where the refundable credit phases in for lower incomes, and (2) inflation lowering the real income level where the credit phases out ($75,000 in nominal dollars for a single parent). The credit was expanded considerably in 2018 by the Tax Cuts and Jobs Act (TCJA); however, the dependent exemption was also eliminated by that bill so that the new CTC is better compared to total tax benefits from the CTC plus dependent exemption in earlier years (and I do not use the 2017 and later period due to data limitations).

An implication of these calculations is that the bulk of the CTC for most families comes through the refundable CTC formula of 15% of earnings over a threshold. While it is possible to receive a nonrefundable CTC that offsets taxes on unearned income, the standard deduction and exemptions for most taxpayers reduce taxable income to zero in the range where the CTC phases in, meaning there is no tax to offset. Nonrefundable credits are thus of little value to families with income under $30,000, and most such families receive a CTC that is directly based on earnings.[3] Thus, the CTC functions very similarly to the EITC, providing a direct subsidy to earnings above a low threshold level.

I next illustrate the CTC's theoretical effects on labor supply through the lens of a standard labor–leisure trade-off model, using Taxsim to compute a budget set for consumption (here, income after taxes) and hours of work for a single worker with two children (Figure 2). Since hours of work reduce leisure, indifference curves increase to the upper left of this figure.

**Figure 2. Budget set with EITC and CTC for a single worker with two children in 2016**

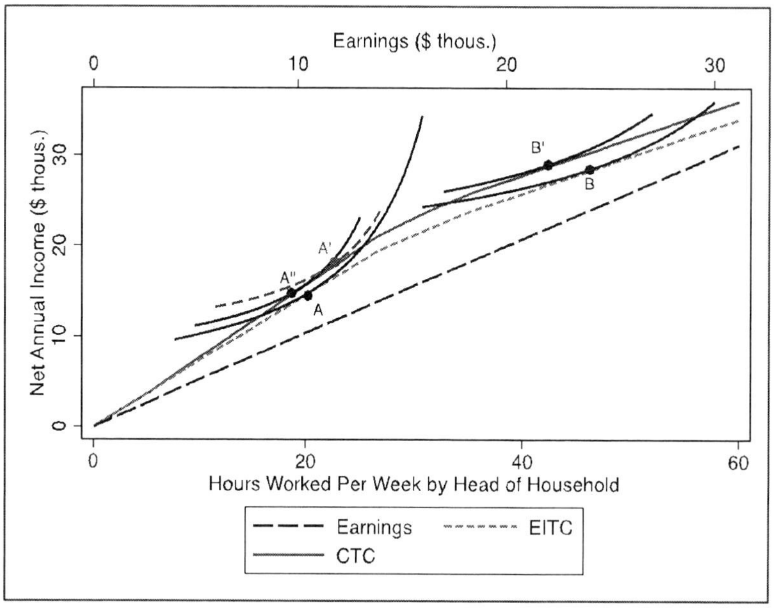


Source: Taxsim calculations for a single parent with two children, working for 52 weeks at $10.00 per hour.

The EITC generates a "hump" on the linear budget set (where net income equals earnings), first subsidizing income in the phase-in region, then remaining constant as a plateau, and finally phasing out with additional hours of work. The CTC augments the effects of the EITC by adding a new wage subsidy above $3,000 in earnings, which continues phasing in until about 31 hours of work per week. At this point, the CTC becomes a lump sum transfer; because the CTC does not phase out until a high income level, a low-wage worker sees no increase in marginal tax rates at plausible values of hours worked.

Thus, the CTC effectively increases the subsidy provided by the EITC, providing differential effects on hours worked depending on whether a worker is in the subsidy or plateau region. In the subsidy region, the effect on labor supply is ambiguous because the positive substitution effect toward work could be counterbalanced by a negative income effect; thus, a worker at point $A$ with no CTC could move to indifference curves at points $A'$ or $A''$. If the substitution effect dominates (as studies of the EITC typically find), the worker would move to a choice with higher work hours, such as point $A'$.

However, in the plateau and phase-out regions, the labor supply effect of the CTC is theoretically predicted to be negative. On the plateau, the consumer faces the same wage rate as without a CTC, so experiences a pure income effect, thereby increasing leisure. The consumer should thus shift from an indifference curve at $B$ to a curve like the one at $B'$ and reduce hours of work. This is especially relevant because if a married couple makes sequential decisions about whether to work, the CTC is typically fully phased in when the secondary earner begins to work, and thus only has an income effect for this worker. Thus, the CTC should encourage labor supply by the primary earner in low-income families but have little or negative effects for secondary earners in those families.

Given that previous research on the EITC finds the strongest effects on the extensive margin (i.e., whether a person works), it is important to specifically consider extensive margin labor supply responses to obtain an accurate picture of the CTC's effects. The CTC lowers the average tax rate of working for primary earners but not secondary earners (as it has typically already phased in for secondary earners). As Eissa, Kleven, and Kreiner (2008) discussed, the relevant parameter for extensive margin labor supply decisions is the average tax rate rather than the marginal tax rate. Because the CTC lowers average tax rates on work, it makes work unambiguously more attractive to primary earners.

### *Previous Research on Labor Supply Effects of the CTC and Other Tax Credits*

As noted by Marr et al. (2015), the effects of the CTC by itself are typically not identified in the literature; this occurs in part because the CTC and EITC phase-in over similar income ranges, making it difficult to disentangle their effects. Some recent exceptions are Kang (2021) and Zheng (2020), which used a DiD strategy with CTC expansions and found elasticity values similar to those in this paper.

The most relevant study on the CTC comes from Feldman, Katuščák, and Kawano (2016), which, like this paper, uses the loss of the CTC when children turn 17 as a source of exogenous variation. However, Feldman, Katuščák, and Kawano focused on higher income households, for whom the loss of the CTC is a lump sum change in taxes and should generate only an income effect on labor supply. They found that loss of the CTC leads to a 0.5 percentage point reduction in earnings growth for the *following* year (an implied intensive margin elasticity of 0.3) but no change in earnings in the year for which credit eligibility is lost. They interpret this effect as taxpayer "confusion" between marginal and average tax rates.[4] While this is suggestive of no intensive margin response to the loss of the CTC, their paper examines only married taxpayers with previous-year income from $30,000 to $100,000, who are less likely to be on the extensive margin. My study therefore expands on Feldman, Katuščák, and Kawano's by including lower-income taxpayers and studying extensive margin changes, where EITC studies have shown stronger effects. Because extensive margin responses are less subject to friction (Chetty 2012), it is plausible that the intensive margin elasticities found by Feldman, Katuščák, and Kawano will be a lower bound for the extensive margin effect of the CTC in my work.

Previous research on the EITC has typically found strong labor supply elasticities, around 0.7 on the extensive margin and 0.1 on the intensive margin (Hotz and Scholz 2003; Nichols and Rothstein 2016). However, these estimates are typically identified from DiD approaches, which rely on unanticipated, permanent changes to the tax system [and thus identify steady state elasticities, in the terminology of Chetty et al. (2012)]. Because responses to temporary changes (known as intertemporal substitution elasticities) are bounded below by responses to permanent changes (Browning 2005), the results from this study may be expected to be higher than the EITC results.

A recent strand of research has questioned some of the earlier findings regarding the strong extensive margin effects of the EITC. Chetty, Friedman, and Saez (2013) found a smaller effect using a more plausibly exogenous identification strategy based on differences in knowledge across neighborhoods; their extensive margin elasticity is around 0.19 (but 0.6 in the top decile of knowledge about the EITC). Kleven (2021) criticized the inconsistent responses to the EITC in DiD studies and provided evidence that the extensive margin elasticity is closer to zero. More generally, Heim (2007) and Kumar and Liang (2016) found declining population-wide labor supply elasticities over time; these may be related to the increasing participation of women in the labor force and the effects of welfare reform, both of which could leave fewer workers on the extensive margin. This study assesses whether the effects of the CTC in a later time period and using a different identification strategy are of a similar magnitude to those found for the EITC.

## Method

### *Data*

I used the 1984 to 2014 panels of the SIPP, focusing on the period 2001 through 2016, where the refundable CTC was in effect but before the TCJA changed the CTC's structure.[5] The SIPP is ideal for my purposes because it provides a large nationally representative panel of US households, is available both before and after the CTC was enacted, and provides information on month of birth for children to construct the running variable in an RD design.

I generated information on tax liabilities and the amount of CTC received by each tax unit using the National Bureau of Economic Research's Taxsim calculator (Feenberg and Coutts 1993), subject to several assumptions detailed in Lippold (2019). My main income measure is post-tax income (which includes labor income, unearned income, taxes, and cash transfers from the SIPP), but I also use adjusted gross income from the tax code [adjusted gross income (AGI), which excludes most cash transfers and taxes] to identify households most subsidized by the credit.

I defined the broad analysis sample as all children who are dependents of a tax unit and were of ages within six months of turning age 14 through 17 at the end of the year. Most households with children have incomes that place them beyond the phase-in region of the CTC, where labor supply effects should theoretically be strongest. Thus, to focus on families who are likely making extensive margin decisions, I exploited the panel nature of the data and defined my primary sample as children with previous-year parental AGI of $0 to $20,000 in 2016 dollars (roughly the bottom 25% of all households with children ages 14 through17). While current-year income is endogenous to CTC eligibility, the use of previous-year income helps isolate low-income households using a predetermined characteristic. This subsample consists of families who faced the greatest incentive to work due to the CTC, and it functions similarly to the restrictions to single parents or workers with low education in studies of the EITC's effects (Eissa and Hoynes 2004; Grogger 2003; Kleven 2021; Meyer and Rosenbaum 2001). The primary sample is a disadvantaged group of parents; 73% are not married, 27% have less than a high school degree (and only 8% have a college degree), and 61% are racial minorities.

I focus on whether either of a child's parents are employed during the calendar year, as measured by an indicator for positive labor income (wages or self-employment, which are the income sources subsidized by the CTC). This measure is comparable to previous studies of the effects of tax credits on labor supply (Eissa and Hoynes 2004; Eissa and Liebman 1996; Meyer and Rosenbaum 2001), which typically use indicators for any work last year.

The unit of analysis for this study is a child because each child's age determines his or her parents' eligibility for the CTC subsidy. A child's tax unit can include either two adults (if parents are married) or one adult (if the child lives with only one parent or guardian, or if the parents are unmarried). My measures of outcomes examined the behavior for the "first mover" in each tax unit (i.e., the presence of any employment or participation), following the analysis of Figure 2, which suggests secondary earners will typically not face greater incentives to work due to the CTC.

### *Identification Strategy*

To investigate the effects of the CTC on labor supply, I exploited the age cutoff for the credit. A convenient feature of the CTC for research purposes is that it provides a credit for each child in the tax unit age16 or below; thus, a child who turns 17 on January 1 of a year will entitle his or her parents to a credit for the previous year, but an otherwise identical child born one day earlier, on December 31, would not provide a credit for that year. Thus, performing an RD using dates of birth for families with children near age 17 can provide internally valid estimates of the effects of the CTC on parental employment for the population of families with older children.

The age 17 cutoff is unrelated to all the major US tax and transfer benefits available to families with children, which sunset when a child turns 18 or 19.[6] Additionally, other changes in children's living situations (such as leaving home or graduating from high school) tend to change at ages other than 17. This implies that changes at age 17 are likely to reflect the CTC rather than the loss of transfer programs or life-cycle factors.

A natural concern about the use of an age cutoff is the possibility of parents manipulating birth timing to gain eligibility for the CTC. Previous research has established that some parents *do* alter the timing of their births around date cutoffs to gain earlier eligibility for tax credits (Dickert-Conlin and Chandra 1999; Gans and Leigh 2009; Schulkind and Shapiro 2014). If parents are able to change when their children are born using elective cesarean sections or induced labor, this would imply that parents can manipulate their eligibility for the CTC, meaning that births would nonrandomly fall in December or January and thus potentially bias estimates of the CTC's effect. A study by LaLumia, Sallee, and Turner (2015), using all income tax returns in the United States, suggests that birth timing effects are indeed present but small in magnitude; they estimate that $1,000 of child tax benefits (including the CTC) are associated with a 1 percentage point shift of births from early January to late December.

In addition to concerns about the manipulation of birth timing, the RD results could be influenced by seasonal trends in birth outcomes. A large body of literature has noted that children born in the winter tend to have worse later economic outcomes than children born in the summer (Bound and Jaeger 1996; Schulkind and Shapiro 2014). For example, Buckles and Hungerman (2013) found that while children born to parents who were not planning to have children displayed no seasonal patterns in parental characteristics, parents who wanted to have children disproportionately gave birth in the spring and summer. To the extent that this makes parents not demographically similar across the December/January cutoff, the RD results will be biased.

These seasonal effects are compounded in my data because the SIPP only includes the date of birth at the monthly level, hiding daily variation that could show more smoothness across the cutoff. Furthermore, for most of my sample period, December children receive one year less of exposure to the CTC than January children (since the credit was introduced after their birth, and the December children age out sooner). This means that the two groups differ in terms of lifetime wealth.

Some evidence for these biases is apparent in Figure 3, which displays the results of this study's RD design among children in the age ranges *before* the CTC cutoff at age 17. For example, an age cutoff at age 16 compares children turning 16 in December to those turning 16 in January within a window of six months on either side of the cutoff. In general, low-income parents of children born in December are likelier to be employed than parents of children born in January, even when both are eligible for the CTC. This seasonal pattern is fairly consistent over time; while the results are imprecise due to low sample sizes for each individual age, and any given discontinuity could reflect either seasonal factors or responses to policies at other age cutoffs, the point estimates of discontinuities are more often positive for children aged 2 to 16. The drop in discontinuities at age 17, where parents lose CTC eligibility, represents a break from the trend.

**Figure 3. December to January discontinuities in labor supply at multiple age cutoffs**

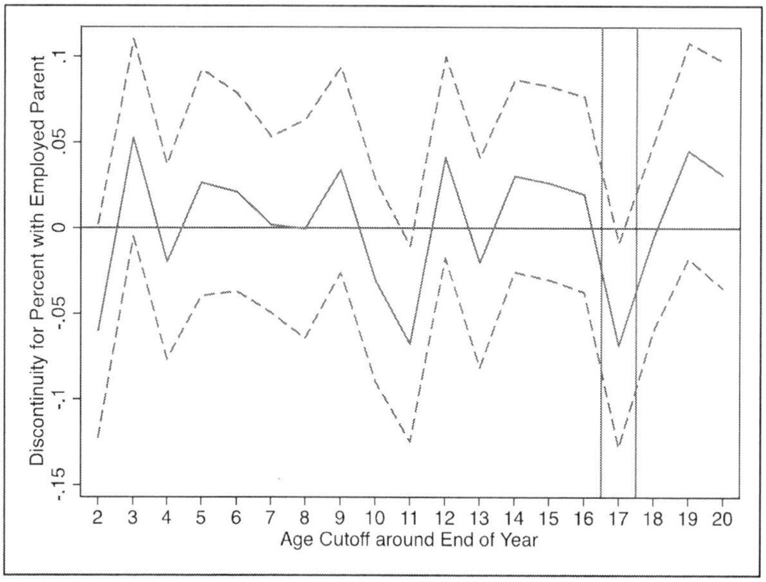


Notes: Discontinuities were estimated in the primary sample using local linear regressions and uniform kernels, with six-month bandwidths around each December age cutoff. The dashed lines represent the 90% confidence intervals. Age 1 was excluded due to imprecision. The results are residuals from regressions with controls (see text).

To address this consistent difference in characteristics, I used a DiRD design (Grembi, Nannicini, and Troiano 2016). While there may be seasonal differences in demographics (observed and unobserved) between parents of children born in different months, those factors should be constant over time. The estimated discontinuity for children at ages under 17 can thus identify the seasonal and birth timing effects, which are then differenced out from the CTC's effect at age 17. Under the assumptions that the seasonal differences are constant and that the effect of the CTC does not itself depend on seasonality—that is, parents with children born in different months respond similarly to the incentives, as discussed by Grembi, Nannicini, and Troiano, the DiRD design will identify the causal effect of the loss of the CTC at age 17 on labor supply.

## *Empirical Model*

My reduced-form DiRD specification is given in Equation (1). I defined $y_{it}$ as an indicator for parental employment for the tax unit of child $i$ for year $t$; the model is thus a linear probability model. The running variable $A_{it}$, capturing age, is the birth month of each child relative to December of year $t$ (normalized so that the December cutoff is zero, November is +1, January is –1, and so forth, until June is –6 and July is +5):

$$y_{it} = \alpha + \delta D_{it} + \phi_1 A_{it} + \phi_2 A_{it} D_{it} + (\gamma + \beta D_{it} + \phi_3 A_{it} + \phi_4 A_{it} D_{it}) T_{it} + \theta X_{it} + \epsilon_{it} \quad (1)$$

As is standard for an RD design, I defined $D_{it}$ as an indicator for the discontinuity (equal to 1 if a child is born in December through July) and included both $D_{it}$ and interactions of $D_{it}$ with $A_{it}$ to estimate the slopes of the running variable on both sides of the cutoff. The baseline coefficient $\delta$ captures the discontinuity at the December–January birth month cutoff for ages prior to the loss of the CTC (estimated by pooling among all the ages before 16.5 included within the regression).

I then incorporated the "difference" aspect of the design by interacting the RD variables with $T_{it}$, an indicator for the child being age 16.5 or above. Thus, the coefficient of interest $\beta$ indicates the jump in outcomes when the child turns 17 in December of year $t$, relative to the jump in outcomes for December births at earlier ages. The interactions with functions of $A_{it}$ allow changes in the slopes of the running variable around age 17 as well.

Equation (1) includes additional variables $X_{it}$ beyond a standard RD framework to increase precision. I included the lagged dependent variable $y_{i(t-1)}$ and other covariates $X_{it}$ in most specifications—namely, year and state fixed effects, parental race (indicator for not White), education (five categories), age (of the oldest parent, quadratic), number of dependents, indicators for married parents, residence in a metropolitan area, and the current and lagged number of months observed. I used the SIPP monthly weights for December in my analysis and followed Nielsen et al. (2009) and Seay and Nielsen (2012) in clustering standard errors for my results by SIPP panel, variance stratum code, and half-sample code to account for the clustered survey design.

A key element of an RD design is the choice of the estimation window. In this paper, there are two windows to consider: the bandwidth for the RD around each cutoff, and the number of age cutoffs to pool when estimating the baseline seasonal discontinuity $\delta$. Because of the cyclical nature of the data, where an age $A_{it}$ value that is far to the right of one December cutoff is inherently located to the left of the next December cutoff, my bandwidth cannot exceed six months without having overlapping observations (preventing separate estimation of linear trends). I chose a six-month bandwidth as the default to maximize power, but consider alternatives in robustness checks. I use the age 14 to 16 cutoffs as "pre-treatment" ages, as these children are around high school age and past the age 13 discontinuity in eligibility for a different tax policy (the Child and Dependent Care Tax Credit); in Lippold (2019), I showed that results are robust to different age choices.

I also created an intent-to-treat (ITT) estimate with a fuzzy RD design (Lee and Lemieux 2010) to directly link the effects of the CTC to the change in outcomes at age 17. To do so, I estimated a first-stage coefficient $\beta_c$ in Equation (1) by replacing the dependent variable with $C_{it}$, the maximum potential value of the CTC received by the parents of child $i$ as of December of year $t$. This CTC maximum value depends only on the number of CTC-eligible children in the family, and will thus mechanically fall at the cutoff due to the child aging out of eligibility. I then simultaneously estimated the reduced form and first stage using seemingly unrelated regression. The resulting ratio of the two discontinuities, $\beta/\beta_c$, estimates the change in probability of employment per dollar of CTC eligibility lost at the cutoff, which I used to construct an elasticity comparable to other studies.

## Results

### *Primary Results*

I first present graphical results illustrating the presence of discontinuities at the December age cutoffs for the period of interest (2001 through 2016, when the CTC was refundable, but before the TCJA changed its structure). Each point in these graphs represents outcomes for children born in a given month, the lowest level of birthdate aggregation available in the SIPP.

I first show that the maximum CTC does drop at the cutoff, as expected (Figure 4). Because Taxsim automatically computes the value of the CTC using the number of children under 17 in each tax unit, the large discontinuity seen here is a mechanical result of the tax model. However, it is reassuring that Taxsim generates a significant drop in the amount of CTC eligibility at the age 17 cutoff, with no discernible discontinuity in the credit received around ages 14 through 16. (Note that the value of the credit does not drop to zero at age 17 because many parents with a child turning 17 have other younger children who remain eligible for the credit).

**Figure 4. First-stage RD results for CTC eligibility of child's parents**

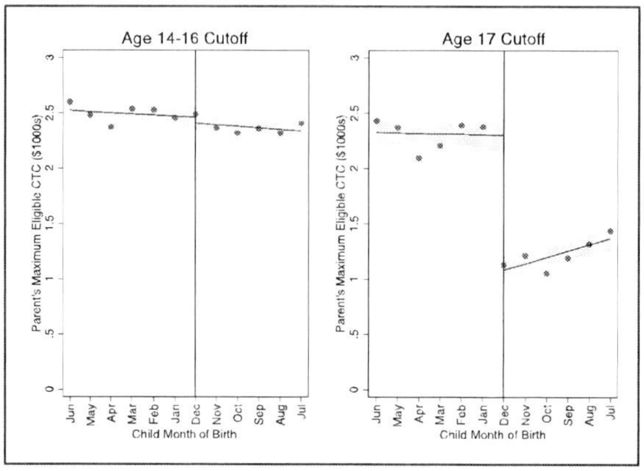


Notes: Discontinuities are estimated in the primary sample using local linear regressions and uniform kernels, with six-month bandwidths around each December age cutoff. The shaded areas represent 90% confidence intervals.

Turning to the primary reduced-form results, Figure 5 presents a standard linear RD design. The data indicate a decline in the proportion of children at the age 17 cutoff in tax units with working parents (although the results are imprecise). However, the discontinuity at ages 14 through 16 is positive, suggesting that seasonal factors (which lead parents with December children to have generally higher levels of employment than parents of January children) partially offset the drop in outcomes at age 17.

**Figure 5. RD results for parental employment**

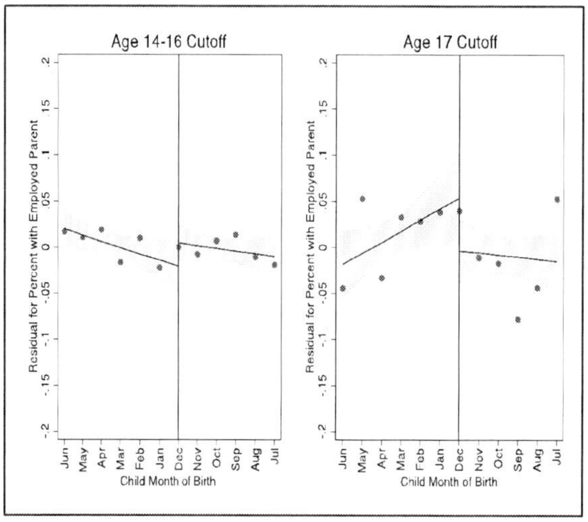


Notes: Discontinuities are estimated in the primary sample with local linear regressions, uniform kernels, in six-month bandwidths around each December age cutoff. Shaded areas are 90% confidence intervals. Results are residuals from regressions with controls (see text).

Figure 6 displays the effect of accounting for seasonality via the DiRD design. Each point in the figure represents the difference between the results estimated around the age 17 cutoff and the pooled cutoff for ages 14 through 16.

The difference in discontinuities at age 17 is higher in magnitude than the age 17 discontinuity alone; estimating the results with controls reduces the variance with little change to the point estimate.

**Figure 6. DiRD results for parental employment**

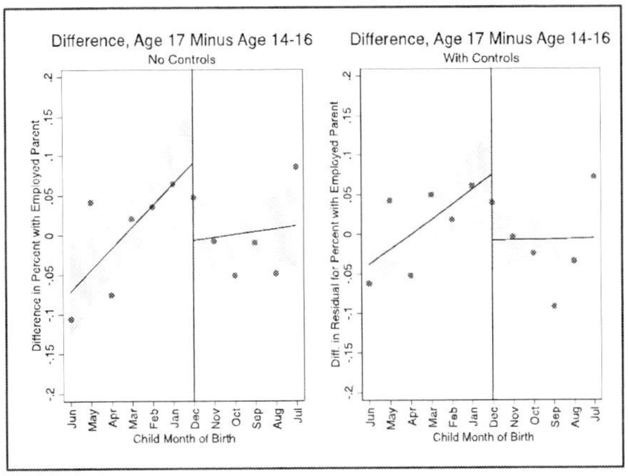


Notes: Points are differences in mean outcomes by month of birth between children around age 17 cutoff (ages 16.5 to 17.5) versus children at the ages 14 through 16 cutoffs (ages 13.5 through16.5). Other notes are as in Figure 5.

These reduced-form DiRD regression results appear in Table 1. The table shows the discontinuity for December for the "pre-policy" ages of 14-16, the average change in outcomes for children around the age 17 cutoff (the "post-policy" period in this context for the DiRD model), and the change in the discontinuities at age 17. In the simplest specification of column (1), the results are large and significant, and remain nearly identical when adding further controls. Column (3), my preferred specification, suggests an 8.4 percentage point fall in the proportion of parents working at the cutoff relative to the 2.6 percentage point seasonal difference between December and January at earlier ages. This result is significant at the 5 percent level.

I also examine the implications of these findings for labor supply in the full population, not just those limited to the primary sample of parents with incomes below $20,000. I found no effect on employment or labor force participation among households with AGI above $20,000. It is reassuring that the effect of the CTC on labor supply appears limited to households where the credit should provide the largest change in average taxes, and thus the largest subsidy encouraging employment.

**Table 1. DiRD results for parental employment by previous-year income**

| Outcome: Working parent | Primary sample: <$20k income (1) | (2) | (3) | $20k–$30k (4) | $30k+ (5) |
|---|---|---|---|---|---|
| Change in RD | −0.097* | −0.087* | −0.084** | 0.049 | 0.001 |
| | (0.054) | (0.044) | (0.042) | (0.032) | (0.007) |
| Age 17 cutoff | 0.091** | 0.067* | 0.076** | −0.022 | −0.003 |
| | (0.044) | (0.038) | (0.035) | (0.026) | (0.005) |
| December RD | −0.007 | 0.021 | 0.026 | −0.002 | −0.001 |
| | (0.036) | (0.021) | (0.021) | (0.017) | (0.003) |
| Lagged outcome | | Yes | Yes | Yes | Yes |
| Controls | | | Yes | Yes | Yes |
| N | 9,443 | 9,443 | 9,443 | 3,988 | 28,188 |
| Clusters | 1,034 | 1,034 | 1,034 | 897 | 1,130 |

Notes: *$p < 0.10$, **$p < 0.05$, ***$p < 0.01$. Standard errors (clustered by variance strata) in parentheses. Sample is 13.5- to 17.5-year-old children in tax units classified by previous-year AGI. Discontinuities estimated with local linear regressions, uniform kernels, in six-month windows centered around December age cutoff. Controls are year and state fixed effects and parental characteristics (see text).

Next, I examined the results by subgroups. I first tested whether the results vary by whether households are entering or exiting employment; while theoretically both types of households face the same potential tax schedule, if there are state-dependent costs to finding employment, households with different levels of labor force attachment could respond differently to the CTC. Table 2 shows the results for the subgroups of households that were not working in the previous year (potential labor force entrants), as opposed to those that were already working (potentially exiting). While the results are not as precise for these subgroups, the estimates are 50% smaller for those previously employed. This implies that the loss of the CTC operates more strongly by reducing the incentive for non-workers to begin work rather than leading current workers to stop working, and that one channel of the CTC's effects is to encourage households to move from nonparticipation into job search and employment.

Table 2 also shows the results by marital status and number of children in a tax unit. The effects of the CTC are fairly consistent across household compositions (conditional on having low previous-year income). The point estimates are slightly lower (but noisier) for households with one child compared to two or more children, which could reflect higher fixed costs in families with more children that leave them closer to the margin of entering employment.

**Table 2. DiRD results for parental employment by subgroups**

| | Previous-year employment | | Marital status | | Family size | |
|---|---|---|---|---|---|---|
| Outcome: Working parent | (1) No work | (2) Work | (3) Single | (4) Married | (5) 1 child | (6) 2+ children |
| Change in RD | −0.108 | −0.046 | −0.095* | −0.064 | −0.041 | −0.101** |
| | (0.080) | (0.041) | (0.052) | (0.062) | (0.072) | (0.051) |
| Age 17 cutoff | 0.117* | 0.037 | 0.101** | 0.012 | 0.070 | 0.075* |
| | (0.066) | (0.033) | (0.044) | (0.050) | (0.050) | (0.044) |
| December RD | 0.026 | 0.005 | 0.015 | 0.032 | −0.000 | 0.031 |
| | (0.042) | (0.022) | (0.025) | (0.041) | (0.044) | (0.024) |
| N | 3,695 | 5,748 | 6,862 | 2,581 | 2,168 | 7,275 |
| Clusters | 810 | 931 | 951 | 689 | 716 | 959 |

*Notes:* *$p < 0.10$, **$p < 0.05$, ***$p < 0.01$. Standard errors (clustered by variance strata) in parentheses. Sample is 13.5 to 17.5-year-old children in tax units with previous-year AGI below $20,000. Discontinuities estimated with local linear regressions, uniform kernels, in six-month windows centered around December age cutoff. Results include lagged outcomes (except columns 1 and 2) and the same controls as Table 1.

## *Robustness Checks*

For any RD design, it is critical to check the robustness of the results, as there are typically many researcher degrees of freedom (Lee and Lemieux 2010). I first tested whether the discontinuity in earnings at age 17 is not spurious by examining whether other demographic variables show jumps at the cutoff (Table 3). In Lippold (2019), I found no significant discontinuity in terms of employment parents' education, race, marital status, age (of the oldest parent), the number of dependents claimed, or in parental or labor force participation in the previous year (which provides a check that the discontinuity does not derive from preexisting differences). To maximize the power of these tests, I test for differences in a standardized index of all of these variables in column (1). This column shows that there are preexisting differences in December at earlier ages, confirming that the DiRD design is necessary, but there is no change in this difference at age 17.

Table 3 also displays the sensitivity of the results to different RD estimation methods in columns (2) to (4), including a triangular kernel, a quadratic polynomial, and a simple average across quarters of the year (effectively, a DiD model at a quarterly level). The results are fairly noisy, but all estimates have the same sign and are similar in magnitude to the baseline in Table 1 (except for the quadratic specification, which has the largest confidence interval).

I also performed a placebo test by examining results from years before the CTC subsidized labor supply. Using the 1984 to 1996 panels of the SIPP, I extend the analysis in my primary sample to the 1980s and 1990s periods, when the CTC did not exist or was effectively not refundable. The results in column (5) show that the estimated discontinuity in this period is insignificant and positive. This implies that some factor changed between the 1990s and the 2000s that discouraged parents with age 17 children born in December from entering the labor force; the CTC provides a plausible explanation for this factor.

Finally, column (6) shows the results that drop states where children can leave school at age 17, as this provides a potential alternative factor that changes at the age 17 cutoff.[7] These results are similar to the baseline specification.

**Table 3. Robustness checks for DiRD results**

| Outcome: Working parent (except (1)) | Outcome: Index of covariates (1) | Triangular kernel (2) | Quadratic fit (3) | Quarterly DiD (4) | Placebo: 1984–1999 (5) | School-leaving age ≠ 17 (6) |
|---|---|---|---|---|---|---|
| Change in RD | –0.071 | –0.049 | –0.007 | –0.037 | 0.044 | –0.091* |
| | (0.132) | (0.048) | (0.075) | (0.026) | (0.065) | (0.047) |
| Age 17 cutoff | 0.064 | 0.073* | 0.064 | 0.043** | –0.092* | 0.069* |
| | (0.123) | (0.041) | (0.071) | (0.020) | (0.053) | (0.038) |
| December RD | –0.137* | 0.022 | 0.023 | 0.009 | –0.043 | 0.034 |
| | (0.072) | (0.023) | (0.034) | (0.014) | (0.031) | (0.023) |
| N | 9,443 | 9,443 | 9,443 | 4,840 | 7,466 | 7,659 |
| Clusters | 1,034 | 1,034 | 1,034 | 899 | 1,395 | 963 |
| Degree | 1 | 1 | 2 | 0 | 1 | 1 |
| Kernel | Uniform | Triangular | Uniform | Uniform | Uniform | Uniform |
| Bandwidth | 6 | 6 | 6 | 3 | 6 | 6 |

*Notes:* *$p < 0.10$, ** $p < 0.05$, *** $p < 0.01$. Standard errors (clustered by variance strata) in parentheses. Sample is 13.5 to 17.5-year-old children in tax units with previous-year AGI below $20,000. Results are local linear regressions with varying parameters and windows centered around December age cutoff. Columns (2) to (6) include the same lagged outcomes and controls as Table 1.

## *Elasticity Estimate*

I estimated the labor supply elasticity (Table 4). Row A provides the previously computed reduced-form estimate for the drop in outcomes at the age 17 cutoff, while row B shows the corresponding first-stage drop in the maximum eligible CTC. The ratio of these two values gives an ITT estimate of an 8.0 percentage point increase in employment for every increase of $1,000 in CTC eligibility.

**Table 4:. ITT and elasticity estimates**

| Outcome | Estimate | Standard Error |
|---|---|---|
| (A) Working parent (change in RD) | –0.084** | (0.042) |
| (B) Maximum eligible CTC (change in RD, $1,000s) | –1.049*** | (0.065) |
| (C) Percent working (lagged mean) | 0.608*** | (0.009) |
| (D) Return to work (lagged mean, $1,000s) | 7.928*** | (0.396) |
| ITT estimate (= $A/B$) | 0.080* | (0.041) |
| Elasticity at average (= $(A/C) \div (B/D)$) | 1.040* | (0.539) |
| N | 9,443 | |
| Clusters | 1,034 | |

*Notes:* *$p < 0.10$, **$p < 0.05$, ***$p < 0.01$. Standard errors (clustered by variance strata) in parentheses. Fuzzy RD and elasticity standard errors are computed using the delta method. Sample is 13.5- to 17.5-year-old children in tax units with previous-year AGI below $20,000. Return to work is difference in post-tax income between working and nonworking parents. Dollar values are in 2016 dollars. Discontinuities estimated with local linear regressions, uniform kernels, in six-month windows centered around December age cutoff. Results include lagged employment and controls (see text).

Next, I rescaled the ITT estimates to compute overall elasticities with respect to the change in return to employment caused by the CTC. To compute the return to work (row D), I calculated the total post-tax income of workers and non-workers in the primary sample, taking the difference in means for lagged income as representative of the expected income growth for a given member of this population when beginning to work. Combined with figures on the mean proportion of parents working in each group (row C), I generated an elasticity estimate at the average, according to the standard formula:

$$\varepsilon = \frac{\Delta y_{it}}{\Delta R_{it}} \cdot \frac{E(R_{it})}{E(y_{it})} = \frac{\beta}{\beta_c} \cdot \frac{E(R_{it})}{E(y_{it})}$$

where $R_{it}$ is the return to work for a given child's parents at time $t$, and the change in the CTC ($\Delta C_{it} = \beta_c$) is the only change in returns to work at the cutoff.

The resulting elasticity estimate is 1.04 in the primary sample of parents with previous-year incomes below $20,000. This is higher than elasticities found in other studies of the effects of changes in CTC-eligible children or dependents on the intensive margin (Feldman, Katuščák, and Kawano 2016; Looney and Singhal 2006). This confirms the theoretical prediction that the CTC would have stronger extensive than intensive margin effects.

However, one reason for the high elasticity is that this study identifies an intertemporal substitution response, as the loss of the credit is temporary (relative to the control group of children born in January who will lose the credit next year) and can be anticipated. Theory thus predicts stronger labor supply responses in this context due to a lack of wealth effects (Browning, 2005). To derive a steady-state measure comparable to studies on the EITC, it is useful to perform a calibration exercise. Using the results derived in Lippold (2019), I obtained the formula:

$$\varepsilon_I \approx \left( \frac{1 - \frac{\gamma W_t}{1 - s_t}\left(1 - \frac{2\alpha}{1 + r_t} + \frac{(2 + r_t)\alpha^2}{(1 + r_t)^2}\right)}{1 - \frac{\gamma W_t}{1 - s_t}} \right) \epsilon_S$$

where the relationship depends on the coefficient of relative risk aversion $\gamma$, the marginal propensity to save $\alpha$, the interest rate on assets $r_t$, the savings rate $s_t$, and the percent change in post-tax income when working $W_t$. In Lippold (2019), I calibrate the model and found that my intertemporal substitution elasticity is consistent with a steady-state elasticities of 0.43 for employment. This value is in the range of results found for the EITC from earlier DiD studies (e.g., Eissa and Liebman 1996 and Meyer and Rosenbaum 2001, as corrected by Chetty et al. (2012), as well as studies using other identification strategies (Chetty, Friedman, and Saez 2013).

## Conclusion

The results presented here suggest that the CTC increases the labor supply among low-income parents of older children, with an estimated steady-state employment elasticity of 0.43. This implies that the original proponents of the refundable CTC were correct and that the credit would increase incentives to work (Sawhill and Thomas 2001). Although some studies have found labor supply elasticities that are lower in the 2000s than in the 1990s, or cast doubt on the EITC's extensive margin effects, the CTC has labor supply effects comparable to the EITC. Thus, my results confirm that tax credits continue to play an important role in encouraging individuals to enter the labor force.

One limitation of these results is that the power of the SIPP sample is not large despite using 16 years of data. This limits the range of hypotheses that could be tested in my sample. Thus, a primary topic for future research is to examine whether the effects shown here replicate in other datasets, such as population-level tax data. Such cross-validation would help confirm my finding that the CTC encourages labor supply.

## Acknowledgments

Thanks to Emily Anderer, Julie Cullen, Gordon Dahl, Alex Gelber, Itzik Fadlon, Roger Gordon, Lane Kenworthy, Craig McIntosh, Katherine Meckel, Stan Veuger, and participants at the 2018 National Tax Association Annual Meeting, 2020 OTA Research Conference, and 2022 FCSM Conference for comments on earlier drafts. This paper was supported by Graduate Summer Research funds from the University of California, San Diego Department of Economics. The views expressed in this paper are those of the author and do not necessarily reflect the views of the US Department of the Treasury.

## References

Acs, Gregory, and Kevin Werner. 2021. "How a Permanent Expansion of the Child Tax Credit Could Affect Poverty." Washington, DC: Urban Institute. https://www.urban.org/research/publication/how-permanent-expansion-child-tax-credit-could-affect-poverty

Bound, John, and David Jaeger. 1996. "On the Validity of Season of Birth as an Instrument in Wage Equations: A Comment on Angrist and Krueger's 'Does compulsory school attendance affect schooling and earnings?'" *NBER Working Paper*, no. 5835. https://doi.org/10.3386/w5835

Browning, Martin. 2005. "A Working Paper from April 1985: Which Demand Elasticities Do We Know and Which Do We Need to Know for Policy Analysis?" *Research in Economics* 59 (4): 293–320. https://doi.org/10.1016/j.rie.2005.09.001

Buckles, Kasey S., and Daniel M. Hungerman. 2013. "Season of Birth and Later Outcomes: Old Questions, New Answers." *The Review of Economics and Statistics* 95 (3): 711–24. https://doi.org/10.1162/REST_a_00314

Chetty, Raj. 2012. "Bounds on Elasticities With Optimization Frictions: A Synthesis of Micro and Macro Evidence on Labor Supply." *Econometrica* 80 (3): 969–1018. https://doi.org/10.3982/ECTA9043

Chetty, Raj, John N. Friedman, and Emmanuel Saez. 2013. "Using Differences in Knowledge across Neighborhoods to Uncover the Impacts of the EITC on Earnings." *American Economic Review* 103 (7): 2683–2721. https://doi.org/10.1257/aer.103.7.2683

Chetty, Raj, Adam Guren, Day Manoli, and Andrea Weber. 2012. "Does Indivisible Labor Explain the Difference between Micro and Macro Elasticities? A Meta-Analysis of Extensive Margin Elasticities." *NBER Macroeconomics Annual* 27 (1): 1–56. https://doi.org/10.1086/669170

Corinth, Kevin, Bruce Meyer, Matthew Stadnicki, and Derek Wu. 2021. "The Anti-Poverty, Targeting, and Labor Supply Effects of Replacing a Child Tax Credit with a Child Allowance." *NBER Working Paper*, no. 29366. https://doi.org/10.3386/w29366

Crandall-Hollick, Margot L. 2018. "The Child Tax Credit: Legislative History." Report R45124. Washington, DC: Congressional Research Service. https://crsreports.congress.gov/product/pdf/R/R45124

Dahl, Gordon B., and Lance Lochner. 2012. "The Impact of Family Income on Child Achievement: Evidence from the Earned Income Tax Credit." *American Economic Review* 102 (5): 1927–56. https://doi.org/10.1257/aer.102.5.1927

Dickert-Conlin, Stacy, and Amitabh Chandra. 1999. "Taxes and the Timing of Births." *Journal of Political Economy* 107 (1): 161. https://doi.org/10.1086/250054

Eissa, Nada, and Hilary W. Hoynes. 2004. "Taxes and the Labor Market Participation of Married Couples: The Earned Income Tax Credit." *Journal of Public Economics* 88: 1931–58. https://doi.org/10.1016/j.jpubeco.2003.09.005

Eissa, Nada, Henrik Jacobsen Kleven, and Claus Thustrup Kreiner. 2008. "Evaluation of Four Tax Reforms in the United States: Labor Supply and Welfare Effects for Single Mothers." *Journal of Public Economics* 92: 795–816. https://doi.org/10.1016/j.jpubeco.2007.08.005

Eissa, Nada, and Jeffrey B. Liebman. 1996. "Labor Supply Response to the Earned Income Tax Credit." *Quarterly Journal of Economics* 111 (2): 605–37. https://doi.org/10.2307/2946689

Esenwein, Gregg A, and Jack Taylor. 1997. "Child Tax Credits: Comparison of Proposals for Low-Income Taxpayers." Report 97-687E. Washington, DC: Congressional Research Service.

Feenberg, Daniel, and Elisabeth Coutts. 1993. "An Introduction to the TAXSIM Model." *Journal of Policy Analysis and Management* 12 (1): 189–94. https://doi.org/10.2307/3325474

Feldman, Naomi E., Peter Katuščák, and Laura Kawano. 2016. "Taxpayer Confusion: Evidence from the Child Tax Credit." *American Economic Review* 106 (3): 807–35. https://doi.org/10.1257/aer.20131189

Gans, Joshua S., and Andrew Leigh. 2009. "Born on the First of July: An (Un)Natural Experiment in Birth Timing." *Journal of Public Economics* 93 (1-2): 246–63. https://doi.org/10.1016/j.jpubeco.2008.07.004

Goldin, Jacob, Elaine Maag, and Katherine Michelmore. 2022. "Estimating the Net Fiscal Cost of a Child Tax Credit Expansion." *Tax Policy and the Economy* 36 (May): 159–95. https://doi.org/10.1086/718953

Grembi, Veronica, Tommaso Nannicini, and Ugo Troiano. 2016. "Do Fiscal Rules Matter?" *American Economic Journal: Applied Economics* 8 (3): 1–30. https://doi.org/10.1257/app.20150076

Grogger, Jeffrey. 2003. "The Effects of Time Limits, the EITC, and Other Policy Changes on Welfare Use, Work, and Income among Female-Headed Families." *Review of Economics and Statistics* 85 (2): 394–408. https://doi.org/10.1162/003465303765299891

Heim, Bradley T. 2007. "The Incredible Shrinking Elasticities." *Journal of Human Resources* 42 (4): 881–918. https://doi.org/10.3368/jhr.XLII.4.881

Hotz, V. Joseph, and John Karl Scholz. 2003. "The Earned Income Tax Credit." In *Means-Tested Transfer Programs in the United States*, edited by Robert A. Moffitt, 141–97. University of Chicago Press. http://www.nber.org/chapters/c10256

Hoynes, Hilary W., Douglas L. Miller, and David Simon. 2015. "Income, the Earned Income Tax Credit, and Infant Health." *American Economic Journal: Economic Policy* 7 (1): 172–211. https://doi.org/10.1257/pol.20120179

Hoynes, Hilary W., and Ankur J. Patel. 2018. "Effective Policy for Reducing Poverty and Inequality?: The Earned Income Tax Credit and the Distribution of Income." *Journal of Human Resources* 53 (4): 859–90. https://doi.org/10.3368/jhr.53.4.1115.7494R1

Kang, Hyein. 2021. "The Child Tax Credit and Labor Market Outcomes of Mothers." University of Kentucky Center for Poverty Research Discussion Paper Series DP2021-09. https://www.ukcpr.org/sites/ukcpr/files/research-pdfs/DP2021-09.pdf

Kleven, Henrik Jacobsen. 2021. "The EITC and the Extensive Margin: A Reappraisal." *NBER Working Paper*, no. 26405. https://doi.org/10.3386/W26405

Kumar, Anil, and Che-Yuan Liang. 2016. "Declining Female Labor Supply Elasticities in the United States and Implications for Tax Policy: Evidence from Panel Data." *National Tax Journal* 69 (3): 481–516. https://doi.org/10.17310/ntj.2016.3.01

LaLumia, Sara, James M. Sallee, and Nicholas Turner. 2015. "New Evidence on Taxes and the Timing of Birth." *American Economic Journal: Economic Policy* 7 (2): 258–93. https://doi.org/10.1257/pol.20130243

Lee, David S., and Thomas Lemieux. 2010. "Regression Discontinuity Designs in Economics." *Journal of Economic Literature* 48 (2): 281–355. https://doi.org/10.1257/jel.48.2.281

Lippold, Kye. 2019. "The Effects of the Child Tax Credit on Labor Supply." Working paper, UC San Diego. https://doi.org/10.2139/ssrn.3543751

Looney, Adam, and Monica Singhal. 2006. "The Effect of Anticipated Tax Changes on Intertemporal Labor Supply and the Realization of Taxable Income." *NBER Working Paper*, no. 12417. https://doi.org/10.3386/w12417

Marr, Chuck, Chye-Ching Huang, Arloc Sherman, and Brandon Debot. 2015. "EITC and Child Tax Credit Promote Work, Reduce Poverty and Support Children's Development, Research Finds." Washington, DC: Center on Budget; Policy Priorities. http://www.cbpp.org/research/eitc-and-child-tax-credit-promote-work-reduce-poverty-and-support-childrens-development

Meyer, Bruce D, and Dan T Rosenbaum. 2001. "Welfare, the Earned Income Tax Credit, and the Labor Supply of Single Mothers." *The Quarterly Journal of Economics* 116 (3): 1063–1114. https://doi.org/10.1162/00335530152466313

National Commission on Children. 1991. "Beyond Rhetoric: A New American Agenda for Children and Families." Washington, DC.

Nichols, Austin, and Jesse Rothstein. 2016. "The Earned Income Tax Credit." In *Economics of Means-Tested Transfer Programs in the United States, Vol. 1*, edited by Robert A. Moffitt, 137–218. November. University of Chicago Press. https://doi.org/10.7758/9781610448789.14

Nielsen, Robert B., Michael Davern, Arthur Jones Jr., and John L. Boies. 2009. "Complex Sample Design Effects and Health Insurance Variance Estimation." *Journal of Consumer Affairs* 43 (2): 346–66. https://doi.org/10.1111/j.1745-6606.2009.01143.x

Sawhill, Isabel, and Adam Thomas. 2001. "A Tax Proposal for Working Families with Children." Welfare Reform and Beyond Policy Brief 3. No. Washington, DC.: Brookings Institution. http://www.brookings.edu/research/papers/2001/01/01childrenfamilies-sawhill

Schulkind, Lisa, and Teny Maghakian Shapiro. 2014. "What a Difference a Day Makes: Quantifying the Effects of Birth Timing Manipulation on Infant Health." *Journal of Health Economics* 33 (1): 139–58. https://doi.org/10.1016/j.jhealeco.2013.11.003

Seay, Martin C., and Robert B. Nielsen. 2012. "SIPP Complex Sample Specification for SAS, Stata, and WesVar." Technical note, Department of Housing and Consumer Economics. Athens, GA: University of Georgia. https://doi.org/10.2139/ssrn.2061581

Urban Institute. 2019. "TRIM3 Project Website." http://trim3.urban.org

Williams, Roberton. 2013. "Who Pays No Income Tax? A 2013 Update." *Tax Notes*, 1615. https://www.urban.org/research/publication/who-pays-no-income-tax-2013-update

Zheng, Wei. 2020. "Child Tax Credit and Maternal Labor Supply." Working Paper, University of Connecticut. https://weizhengecon.wixsite.com/website/research

## Endnotes

1 The original CTC was refundable for families with three or more children, but only to the extent that their payroll taxes exceeded the family's EITC. Accordingly, refundable CTC amounts were only 3% to 5% of total CTC expenditures in the first few years of the credit (author's calculations from Internal Revenue Service data). This alternate payment formula still exists in current law, but as a family with three or more children will almost always have EITC exceeding payroll taxes at low levels of income because the EITC subsidizes earnings at a 34% to 45% rate, whereas payroll taxes are 7.65%. This formula is rarely used and is not a focus in this or other studies of the CTC.

2 Results are similar for (1) married taxpayers, who face the same slopes but begin the phase-out for both the EITC and CTC at a higher income level, and (2) differing numbers of children, which rescales the maximum level of the credits but does not change the slopes. See Lippold (2019) for details.

3 For example, Williams (2013) found that over 99.5% of tax units with children and incomes under $30,000 owed no income tax in 2013. Those figures included the CTC, but my calculations with my 2001–2016 SIPP panel and Taxsim confirmed that 96% of such families owed no income tax in this dataset, and 94.3% of such families would owe no tax even without the nonrefundable CTC. The American Rescue Plan did make the CTC fully refundable for 2021 only, but that is after the time period of this study.

4 Notably, the authors do not discuss the potential role of average tax rates on labor supply; while they interpret their finding as a reduction in labor supply that is irrational Because marginal rates have not changed, the theoretical results of Eissa, Kleven, and Kreiner (2008) showed that it can be optimal to change labor supply in response to a change in *average* tax rates in the presence of fixed costs of work. Feldman, Katuščák, and Kawano's robustness checks also showed that the effect on labor income is strongest for households with incomes in the $30,000 to $55,000 range, which could be consistent with extensive margin responses by a portion of this group. Furthermore, because they examine labor supply in the year *after* CTC eligibility changes, their use of variation in eligibility for the CTC cannot be separated from the effect of children turning 18 in the tax year, which could have effects on labor supply through channels other than the CTC.

5 Because of the SIPP's change to use overlapping samples in the 2018–2020 panels, as well as high nonresponse rates during the COVID-19 pandemic, I leave analysis of years later than 2016 for future work. The SIPP's panel structure does not allow me to analyze year 2000.

6 In particular, the dependent exemption and EITC end at age 19 (or age 24 if the child is a full-time student); the childcare credit and childcare subsidies end at age 13, and children receive special treatment under SSI and SNAP until age 18. TANF rules change at age 16, 18, or 19, depending on the state and child's student status; Medicaid and CHIP end at age 19; and WIC benefits end at age 5. (All parameters taken from Urban Institute 2019.)

7 School-leaving ages are drawn from the Digest of Education Statistics (various years), https://nces.ed.gov/programs/digest.

# V. Firm Power and the Law

## Regulatory Changes and Labor Market Concentration in the US Agriculture Sector: Evidence from H-2A Workers

**Candice Yandam Rivière**
*Judicial Law Clerk, US Court of Appeals for the Second Circuit*

### I. Introduction

For decades, the United States has relied on guestworker programs for low-skilled workers. One of these programs, the H-2A program, is specifically designed to allow farmers and growers to recruit foreign workers seasonally.[1] H-2A workers—seasonal migrant workers who hold an H-2A visa—are authorized to work in the United States for a few months. After this period has elapsed, they must return to their home country.

Some scholars have highlighted the benefits of guestworker programs, as they allow workers to receive higher pay than they otherwise would. These programs also allow them to send part of their earnings to their home country, which contributes to better educational outcomes and increased national income.[2] Host countries benefit from a cheaper workforce, which leads to lower prices. On the flip side, guestworker programs are known for being exploitative and abusive.[3] And, for host countries' workers, an influx of cheaper workforce means less bargaining power and lower wages for everyone.[4]

This paper's main goal is to provide some understanding of the law-and-economics dynamics behind the H-2A program. Some scholarship has been written on the legal dynamics of the H-2A program. It emphasizes the program's failures, such as failing to protect US workers, not allocating visas efficiently, and exploiting foreign workers.[5] Other scholars have looked at the economics of temporary visa programs, including the H-2A program. A recent paper analyzes the effects of labor market monopsony on H-1B, H-2B, and H-2A workers,[6] and the authors find that generally, guestworkers face moderately or highly concentrated labor markets and that it decreases wages. However, to the best of my knowledge, no paper has attempted to analyze the H-2A program's statutory framework, the impact of recent changes in regulation, and how the framework's effects influence labor market power in the agriculture sector. This is precisely what this paper attempts to do.

First, this paper explains that the H-2A programs unsuccessfully balance two seemingly competing goals: to allow migrant workers to fill labor shortages in the United States and to protect domestic workers from the adverse effects of increased labor supply. From an economics perspective, labor supply shocks necessarily create at least some adverse effects for workers in the receiving labor market (the market absorbing the labor supply shock).[7] Additionally, the H-2A program is plagued by fraud and systemic abuse, and contributes to collusive practices, which leads to more concentrated labor markets.

Second, this paper discusses a major regulation in the H-2A program. In 2015, the Department of Labor (DOL) provided that herders, a category of H-2A workers that had previously been left out of the main H-2A regulatory scheme, would be subject to the same protections and the same wages as the other H-2A workers. The difference-in-differences estimates show that the change in regulation decreased the demand for H-2A herders but increased the number of hours worked for those herders compared to other nonherder H-2A workers. While those results present only a partial view of the labor market dynamics at play, they tend to show that the 2015 change had the anticipated effect: by raising the costs of hiring H-2A workers, the demand for those workers dropped, potentially increasing the labor demand for US workers instead.

Third, the paper attempts to show that collusive practices lead to more concentrated labor markets in agriculture. In fact, the data show that labor markets that regularly hire H-2A workers are moderately or highly concentrated depending on the type of job. Descriptive regressions of wages on labor market concentration show that a more concentrated labor market leads to lower wages. Although the magnitude of the results is small, those results are in line with other scholarship on labor market concentration. Additionally, the results are probably conservative because the measure of labor market concentration is generous and might not account for the role of agricultural associations, known for hiring almost all H-2A workers in certain states.

The rest of the paper proceeds as follows. Part II goes over the labor certification requirements for H-2A workers and their legal interpretations by courts. Part III demonstrates that the H-2A program leads to economic inefficiencies and perpetuates collusive and abusive labor market practices, followed by a discussion of the 2015 regulatory changes in Part IV. Then, I present the data in Part V and attempt to determine the effects of a change in the H-2A program regulation and discuss the difference-in-differences estimates in Part VI. Part VII analyzes the role of labor market concentration on migrant workers' wages, and Part VIII discusses the limitations and implications of the results' implications. Part IX addresses policy considerations, which are followed by the conclusion in Part X.

## II. Labor Certification Requirements and Legal Interpretation

The United States has relied on temporary immigration for decades. Certain foreign workers, also referred to as guestworkers, can be admitted to the United States to perform temporary labor with an H-2 visa. Specifically, the H-2A visa category allows foreign workers temporary admission in the United States to perform seasonal or temporary agricultural labor. For example, they engage in range herding, crop harvesting, and livestock production. There is no formal education requirement for these jobs. This section will address the history of the H-2 program, the labor certification regulation and the process as it occurs currently, and the courts' interpretation of the regulation.

### *A. Brief History of the H-2A Program*

The H-2 visa program was created following World War II through a series of bilateral agreements between the United States and other countries. During WWII, the United States imported its workforce from the Bahamas, Jamaica, Barbados, other Caribbean islands, and Mexico. After the war, the United States and foreign countries wanted to keep their arrangements. Other similar arrangements were established, for example, with the Bracero program between the United States and Mexico.[8] During the 1940s and 1950s, thousands of workers were admitted to the United States and performed supplemental seasonal work.[9]

The H-2 category of visas was first introduced by the Immigration and Nationality Act of 1952. The category was primarily designed to import unskilled workforces. Later, the Immigration Reform and Control Act of 1986 subdivided the H-2 category into two subcategories: the H-2 A category, which is an uncapped category of visas for temporary agricultural workers, and the H-2B category, which is a capped category of visas for temporary workers in other domains. In part, at the time of the enactment, the importance given to temporary agricultural workers was intended to counterbalance the anticipated potential decline in the illegal immigrant population due to the increased border security and immigration enforcement provisions of the Act. In Table 1, after a sharp increase in the number of undocumented noncitizens apprehended by US authorities in 1954, there was a steady decline for about a decade, suggesting that the act may have had its intended effect.

**Table 1. Undocumented noncitizens apprehended by US authorities (1940–1976)**

| Year | No. Apprehended | Year | No. Apprehended |
|---|---|---|---|
| 1940 | 10,492 | 1959 | 45336 |
| 1941 | 11,294 | 1960 | 70,684 |
| 1942 | 11,784 | 1961 | 88,823 |
| 1943 | 11,715 | 1962 | 92,758 |
| 1944 | 31,174 | 1963 | 88,712 |
| 1945 | 69,164 | 1964 | 86,597 |
| 1946 | 99,591 | 1965 | 110,371 |
| 1947 | 193,657 | 1966 | 138,520 |
| 1948 | 192,779 | 1967 | 161,608 |
| 1949 | 288,253 | 1968 | 212,057 |
| 1950 | 468,339 | 1969 | 283,557 |
| 1951 | 509,040 | 1970 | 345,353 |
| 1952 | 528,815 | 1971 | 420,126 |
| 1953 | 885,587 | 1972 | 505,949 |
| 1954 | 1,089,583 | 1973 | 655,968 |
| 1955 | 254,096 | 1974 | 788,145 |
| 1956 | 87,696 | 1975 | 766,600 |
| 1957 | 59,918 | 1976 | 875,915 |
| 1958 | 53.474 | | |

Source: Kitty Calavita, Inside the State: The Bracero Program, Immigration, and the I.N.S. 236 (1992) [compiling numbers from the US Immigration and Naturalization Service, *Annual Report* (1959)].

Since then, the H-2 A program has evolved considerably. Today, the goal of the program is to provide an agricultural workforce for employers facing labor shortages. Although the number of visas issued has decreased since the 1950s, the United States still imports a significant number of H-2A workers. Moreover, as Figure 4 in the Appendix shows, this number has consistently increased over the past few decades. Thus, the program is still an important source of labor for the US agriculture sector.

## *B. The Specifics: H-2A Program Administration and Legal Interpretation*

### *1. Administration of the H-2A Program*

The Immigration and Naturalization Act provides that an H-2A petition cannot be approved unless the employer or petitioner has applied for a labor certification to the DOL and that the DOL has certified that

- There are not sufficient workers who are able, willing, qualified (or equally qualified in the case of aliens who are members of the teaching profession or who have exceptional ability in the sciences or the arts), and available at the time of application for a visa and admission to the United States and at the place where the [noncitizen] is to perform such skilled or unskilled labor, and
- the employment of such [noncitizens] will not adversely affect the wages and working conditions of the workers in the United States similarly employed.[10]

Those requirements have a dual goal: to provide a steady labor force to employers in the agriculture sector and to protect US workers' jobs, wages, and general working conditions.[11] Thus, the H-2A program grants foreign workers temporary admission to the United States to "perform agricultural labor or services of a seasonal or temporary nature, provided that US workers are not available."[12]

Generally, H-2A workers are admitted for a period of time that does not extend more than a year. An approved H-2A visa petition is therefore valid for up to one year, but an employer can apply to extend an H-2A worker's stay in the United States for up to one year. However, a noncitizen total period of stay with an H-2A visa cannot exceed three consecutive years. A noncitizen who has spent three years in the United States as an H-2A worker may not seek an extension of stay or be readmitted to the United States under the same visa category until she has been outside the country for three months.[13]

The H-2A program is administered by the Employment and Training Administration (ETA) of the DOL and US Citizenship and Immigration Services (USCIS) of the Department of Homeland Security (DHS). The process for labor certification entails a few steps.[14] First, employers must apply to the DOL for labor certification. Second, after receiving labor certification, a prospective H-2A employer can submit an application, also known as a petition, to DHS to import a foreign migrant workforce. Third, if the application is approved, a foreign worker can go to a US embassy or consulate to apply for an H-2A nonimmigrant visa from the Department of State (DOS). Last, if her visa application is approved, the worker can use this visa to apply for admission to the United States. Table 2 shows that over the past years, there has been an increase in the number of petitions requested and a constant number of petitions certified, despite the fact that there is no formal quota for H-2A visas (unlike H-2B or H-1 visas).[15]

**Table 2. H-2A petitions requested and certified (2015–2019)**

| Actions | 2015 | 2016 | 2017 | 2018 | 2019 |
|---|---|---|---|---|---|
| Positions requested | 368,148 | 398,664 | 403,086 | 418,641 | 420,294 |
| Positions certified | 352,317 | 374,345 | 373,257 | 353,751 | 356,409 |
| Percentage certified | 95.7% | 93.9% | 92.6% | 84.5% | 84.8% |

Source: Data compiled by author from USCIS, Immigration and Citizenship Data.[16]

*2. Court's Interpretation of the Labor Certification Statutory Requirements*

The statute has a dual purpose: to admit noncitizens and to protect US workers from the negative effects of an increased labor supply. By implication, the statute can be understood as protecting US workers from adverse working conditions and lower wages. Such adverse conditions would arise if more workers were to work in the country because an influx of workers would decrease the demand for workers and would negatively impact wages. Court opinions and administrative decisions lend a lot of support to this view.[17] For example, in *Saxbe v. Bustos*,[18] the Supreme Court held that the purpose of the statute was to limit new admissions of noncitizen laborers,[19] while other courts also highlighted the importance of admitting and absorbing foreign workers while protecting US workers.[20] However, despite the fact that courts have recognized this dual goal, the statute has often been interpreted in conflicting and sometimes inconsistent ways.

For instance, courts, in determining whether a labor certification will be granted under the statute, held that the DOL must determine whether "there are not sufficient workers in the United States who are able, willing, qualified, and available at the time of the [noncitizen]'s application, and at the place"[21] where the noncitizen plans to perform the labor in question. This is very important because it allows the DOL to assess the labor market and make sure there are not enough US workers in an area to do the job. If there are not enough US workers, then allowing migrant workers in the labor market will presumably not have a negative effect on US workers. In one instance, the Board of Immigration Appeals held that a noncitizen had received an invalid labor certification under § 1182(a)(14) because the certification reflected that the noncitizen was to be paid $1.50 per hour for animal skinning, when it should have been paid the prevailing rate of $3.00 per hour.[22] Further, the board stated that under the circumstances, the actual arrangements between the employer and the noncitizen frustrated the requirement of § 1182(a)(14) because the DOL indicated that if the job in question were made available at the higher rate of $3.00 per hour, local workers with experience as animal skinners and butchers would have accepted it. And, noting that two applicants had expressed an interest in the job at that rate, the board concluded that the arrangement in question did not meet the requirement that certifications would be issued when there were not sufficient workers in the United States who were able, willing, qualified, and available to do the labor at the place where the alien was to perform the labor.

The labor certification process under the H-2A program works similarly. For an employer to hire a foreign worker as an H-2A worker, the employer must petition the Secretary of Labor for a temporary labor certification. In one instance, the court held that if US workers did not accept employers' conditions, an employer was legally allowed to hire migrant workers. Once a labor shortage is established under § 1188(a)(1),[23] domestic employees cannot challenge the recruitment of migrant workers on the basis that domestic employees are available to work for a higher wage. In *Hernandez Flecha v. Quiros*,[24] the court held that if workers are not willing or able to work under the conditions required by the employer, then they meet the definition under § 1188(a)(1)(A). Otherwise, the court explains, the statute could be used as a tool by domestic workers to request many advantages from their employers. In turn, "the necessary effect would be that the alien market would never be reached [and] the employer would have to pay whatever the domestic workers sought, it being obvious that if there were no limit on the price that could be asked, workers could always be found."[25]

Second, courts must determine if the Secretary of Labor, when granting or refusing labor certification, properly analyzed whether it would "adversely affect the wages and working conditions of US workers similarly employed" within the meaning of the statute. Again, courts have been inconsistent in their interpretation of this part of the statute. In *Pesikoff v. Secretary of Labor*,[26] the court held that the Secretary's denial of a labor certification for a live-in maid was entirely appropriate. The court found that the Secretary could well predict that the wages and working conditions of US maids would be adversely affected by employers importing foreign workforces. By contrast, in *Elton Orchards, Inc. v. Brennan*,[27] apple orchard owners challenged the Department of Labor's policy of allocating inexperienced domestic apple pickers to one orchard while allowing other farm owners to import experienced apple pickers from the British West Indies, from which other farms (now required to use inexperienced pickers) were used to hire workers. The court ruled that despite recognizing the urgency in the proper harvesting of apples (if they are not promptly harvested, they might get spoiled) and the fact that they had a previous work relationship, the DOL's policy prevailed. In the words of the court: "To recognize a legal right to use alien workers upon a showing of business justification would be to negate the policy which permeates the immigration statutes, that domestic workers rather than aliens be employed wherever possible."[28]

Courts have interpreted the statute inconsistently when dealing with H-2A workers as opposed to other workers. Yet labor certification requirements are the same. This lack of uniformity in interpreting the statute likely stems from two areas: a lack of understanding of fundamental labor market principles and a history of abusing particularly vulnerable migrant workers.

## III. The H-2A Regulation Leads to Economic Inefficiency and Perpetuates Abuses Despite Regular Changes

Despite its seemingly rational dual goals, the statutory requirements of labor certification will necessarily lead to adverse economic consequences for US workers, and as they operate today, create and perpetuate a climate of abuse toward H-2A workers. This section will discuss the direct and indirect consequences of the regulation, as well as the abuse that the regulatory scheme has, in part, contributed to perpetuating.

### *A. Economic Consequences of the Requirements*

The labor certification regulation, as is, leads to direct negative consequences on US workers and practices stemming from the regulation contribute to wage-fixing and colluding practices among employers.

*1. Direct Consequences: Adverse Effects on Workers*

The labor certification requirements under the statute suffer from attempting to reconcile two goals that do not need to be reconciled under basic economic theory. On the one hand, the statute attempts to protect US workers from labor supply shocks that would occur if more migrant workers were admitted to the country. On the other hand, the statute is designed to provide a migrant workforce to employers who are facing labor shortages. Courts have understood these goals to be mutually exclusive and have emphasized the need to "strike a balance between the two goals."[29]

First, there is no need for a statute to "strike a balance" between "assuring an adequate labor force" in the meaning of § 1188 and "protecting the jobs of the citizens"[30] in a competitive labor market because these two goals are not mutually exclusive. In a healthy labor market (free from collusion or other market failures), labor supply and demand are regulated through the competitive wage within the market.[31] The wage will adjust to fluctuations in supply and demand, and an adequate labor force would necessarily protect the jobs of domestic workers.

Second, from an economics perspective, it is difficult to entertain the idea of a labor shortage. Theoretically, if there is no supply of workers at a given wage, then to attract laborers, employers will have to increase the wage.[32] A labor shortage exists only when there is a physical shortage of people (for instance, during the second World War, there was a shortage of men). Of course, at points in time, within a given labor market (a specific geographic zone), there can be labor shortages. Again, though, even if all working individuals are employed, they may still switch jobs to work for a more attractive wage, for example. In other words, a purported "labor shortage" should not be the rationale underlying the statutory scheme, as it is based on a misunderstanding of labor market dynamics.

Third, an increase in labor supply within a market almost always results in some "adverse effect on wages" on domestic workers, at least in the short run.[33] Consider a sudden increase of workers in a market (a labor supply shock). There will be more supply for a specific job for a constant demand, which will lead to a decrease in wages for all workers (United States and foreign). Unless foreign and domestic workers are complements (as opposed to substitutes), they will compete with each other, which will lead to lower wages and adverse effects.[34]

*2. Indirect Consequences: Fraud and Collusion*

As established previously, the first condition to qualify for petitioning to import migrant labor under the H-2A program is to demonstrate that "there are not sufficient workers who are able, willing, and qualified, and who will be available at the time and place needed, to perform the labor or services involved in the petition."[35] The certification process is supervised by the DOL. However, the DOL does not ensure that employers are not artificially creating conditions that lead to a purported labor shortage and would justify hiring H-2A workers.

Before certifying an H-2A petition, the DOL asks the employer to prove the existence of a labor shortage. The employer has to determine the current availabilities in the domestic labor market. In practice, the employer must first seek to recruit domestic workers to fill the required positions. Employers post job vacancies containing the terms and conditions of employment to the DOL-funded state workforce agency (SWA)[36] in the specific area of intended employment *before* an employer can submit an H-2A labor certification application. However, the recruitment process is riddled with deceit and procedural problems.

The purpose of these job postings is to make sure that, despite a thorough and good faith effort to recruit US workers, employers were unsuccessful and are suffering from a labor shortage in their geographic area. In turn, the situation would trigger § 1188(a)(1) and allow employers to import H-2A workers. All employers must go through this step of the recruitment process, but they often signal on job postings that they are specifically recruiting H-2A workers, effectively preventing domestic workers from applying to those jobs. A search of states' databases containing current job listings yields interesting results. Job listings posted on California's Workforce Agency (CalJobs) show that, although harvesting and farming jobs are open to US workers, job postings usually start with "H-2A."[37] A US worker looking for a job as a crop harvester may not consider such a job because they are not an H-2A worker.

Although the process is slightly different for herders, the results are the same. A job search for "herder" tons of job offers, most of them for a season (such as "Range Winter Sheep Herder") and include the mention "H-2A," sometimes in the job title, sometimes in the job description.[38]

The job postings on states' websites are likely not targeting future H-2A workers either. To apply for a job through these websites—that is, to create an online account and apply—one must provide a Social Security number (SSN), effectively blocking foreign workers who do not have a US SSN from even creating an online account.[39] Additionally, employers do not usually rely on state agencies to recruit H-2A workers. Instead, "when USCIS approves an employer's petition, the employer often hires labor recruitment companies to locate workers for prearranged terms of employment."[40] Since labor recruitment companies work directly with workers and employers, it is unlikely that they need to refer to online job postings on state databases; recruitment agencies know what the employers they work with are looking for and presumably have access to a large pool of interested applicants.

Far from uncommon, the H-2A certification process suffers from other procedural fraud. Indeed, other instances of procedural failures have been reported regarding the certification of H-2A applications, such as approvement of deficient applications and failure to properly verify H-2A workers' working conditions.[41] As a result of these bad faith practices and lack of agency oversight, employers can fabricate the conditions of a labor shortage in the domestic market, and that the recruiting of migrant workers would not hurt domestic workers. In turn, they can hire H-2A workers.

Collusion has also been a problem, especially from a statutory perspective. Congress has explicitly permitted agricultural associations to make collective decisions about hiring migrant labor, including decisions about their wages.[42] Petitioners or employers can apply for a certification through an agricultural association. In sum, agricultural associations are allowed to act as sole employers in the context of the H-2A certification.[43]

The main issue with associations acting as employers in the H-2A program is that they have labor market power because they represent large shares of the labor market and allow employers to act in unison with regard to the certification process. For example, in *Llacua v. Western Range Ass'n*,[44] the two associations at issue were responsible for hiring 91% of shepherds. In practical terms, this means that the associations can hire most of the shepherds in a specific region, which gives them considerable labor market power. If agricultural associations of ranchers can decide who they want to hire and at what price, and if they have labor market power, they could easily be "restraining trade." When a rancher, and an association of ranchers (such as the WRA or the MPAS) petition together (co-signing the H-2A visa), to hire a foreign worker, it could be thought of as an agreement between them to fix wages and make collective decisions about hiring.

Even if courts have not recognized associations' behavior in the H-2A context to violate antitrust law, it does not mean that these collusive practices don't exist.[45] In fact, they have existed for a long time, and are deeply rooted in racist stereotypes and cultural narratives that allow for exploitation of the foreign workforce and the undervaluing of the US workforce.

## *B. False Cultural Narratives Based on Racist Stereotypes Mask Systemic Abuse*

Guestworker programs in agriculture are based on pervasive narratives that harm US workers and foreign workers. Initially, the H-2A program was accepted because it was based on the fact that migrant workers were filling the gaps in places where there were labor shortages. However, as early as 1951, growers were portraying US workers as lazy, and that "no amount of money would induce [them] to accept steady employment."[46] Even some senators echoed similar sentiments: A guestworker program was absolutely crucial to the agriculture sector because these jobs "are rejected by our citizens" who "find it more profitable and secure to collect unemployment or public assistance."[47] Of course, such narratives were used to mask growers' interests in maintaining a cheap and steady workflow from abroad.[48] By failing to pay competitive wages and accusing US workers of laziness, growers ensured that the H-2A program would provide them with workers willing to work for the wage they provided.

By contrast, foreign workers were praised for their "natural adaptation to agricultural work."[49] Such narratives were based on racist and eugenic-based stereotypes: Mexican workers were deemed more productive as harvesters because of their "short stature,"[50] while Asian immigrants were praised for being "docile and efficient."[51] In truth, employers were authorized "to shape these undesirable jobs and coerce guestworkers into high rates of productivity."[52] Guestworkers often remained silent, did not demand raises or question working conditions for fear of getting fired and having to return to their home country.

In sum, the societal narratives that justify the existence of these exploitative practices are as follows: labor shortages exist because US workers are lazy and refuse to do these difficult jobs, while guestworkers are deemed as working harder and being more productive.[53] In reality, employers, within this context, might be "part of a calculated effort … to lower labor costs and expand profits,"[54] at the expense of workers.

These cultural narratives are still very much alive today. They resonate with people because they play on racialized notions that are perpetuated throughout society. In agriculture, employers prefer hiring Latino immigrants because of their "essentially unchangeable cultural inclinations or characteristics," which some also understand as making them much less suited for managerial responsibilities.[55] However, white workers are perceived as not wanting "to get their hands dirty" because "they are too good for low-wage jobs."[56]

H-2A workers are also subject to widespread abuse, such as wage theft, fraud, coercion, harassment, stemming from considerable power imbalances.[57] Wage violations are the most common complaints among H-2A workers.[58] Many H-2A workers explained that the salary they received was much less than what they were told they would receive during the recruitment process in Mexico. Regular illegal deductions considerably cut these workers' wages, and "[m]described being paid less than the legal wage—sometime far less."[59] Other extremely serious violations occur, such as sexual violence, verbal abuse, or lack of freedom.[60] These reports suggest that H-2A regulations are poorly enforced and that any regulatory change might not have much of a positive effect on H-2A workers.

## IV. Recent Regulatory Changes

The economic challenges and abuses described in Part II of this paper have spurred regular changes in the administration of the H-2A program. One of the biggest changes over the past decade occurred in 2015, when the DOL issued a new final rule to update the amend the H-2A program to better protect herders.

### *A. The 2015 Change*

Prior to 2015, all H-2A workers were subject to the same regulations except for H-2A workers working in range herding and livestock production. At the time, the rationale was that the standard H-2A rule requirements did not apply to these unique occupations that "place workers in remote locations where they are on call 24 hours per day, seven days a week."[61] Because these occupations were exempted from the same regulatory regime, shepherds were often working under difficult conditions and for as low as $4.5 per hour.[62]

In 2015, the DOL issued regulations for range herding and livestock production jobs that established new requirements for employers applying for labor certification to hire H-2A workers for herding work. Essentially, these standards and procedures include the labor certification process, such as "job order and labor certification application filing, US worker recruitment, wage requirements, and housing standards."[63] The new standards and procedures put nonherder and herder H-2A workers subject to the same requirements.

### *B. Wage Changes*

One of the H-2A program's goals is to prevent adverse effects on similarly employed US workers. To satisfy this goal, the H-2A program controls wage levels and forces employers to offer wages at or above certain levels. For instance, under the H-2A program, employers must pay their workers "the highest of the federal or applicable state minimum wage rate, the applicable prevailing wage rate, the adverse effect wage rate (AEWR), or the agreed-upon collective bargaining wage rate."[64]

One of the major policy debates about the H-2A program revolves around the proper methodology for determining the AEWR.[65] For instance, labor advocates have argued that the AEWR protects US workers from lower wages due to an increase in labor supply resulting from the hiring of foreign workers. Employers, on the other hand, have emphasized that the AEWR results in inflated wage rates that may lead some farmers to declare bankruptcy and workers to lose their jobs.

## V. Data

The data used in this paper are from the Department of Labor Foreign Certification Performance data.[66] The data contain petition-level information on H-2A applications and certifications. Specifically, each observation in the data matches a petitioner (usually a firm) with an H-2A worker. During the H-2A application process, employers first file a labor condition application (LCA) and disclose a number of information about themselves. If the application is accepted, they are allowed to apply for an H-2A visa with USCIS. Because each observation in the data records information about an H-2A petition in a given year, it includes minimal information about the employee and some information about the petitioner.

For the following empirical analyses, I used the years 2014 to 2019. Data are available for 2020, but the COVID-19 pandemic may induce bias in the estimates.[67] The data require some cleanup to be used properly. First, I narrowed my sample to H-2A petitions that have been granted, following the work of Gibbons, Greenman, Norlander, and Sørensen.[68] Petitions that have been denied are not very interesting because the workers will not be participating in the US labor market. Second, I create a variable that counts the number of job vacancies (also called job postings) by a specific firm within a specific labor market to measure market concentration. Many studies in the economic literature have used job vacancies as a measure of firm-specific labor demand and the concentration of job vacancies at particular firms within a labor market as a measure of the concentration of labor demand within that market.[69] Last, I dropped all certified petitions that had expired.[70] I also dropped outlier applications or applications for which no information was disclosed.[71]

For each analysis, a subset of the data were used. Descriptive statistics are provided in Table 4 and Table 6 for the subsamples used in each analysis, and more details on the data are also provided.[72] In each analysis, data limitations will be discussed in some detail.

## VI. Some Effects of Regulatory Changes on the Labor Market

In 2015, as described in the third section of this paper, the H-2A program underwent a major change in its administration and regulation.[73] For shepherds and herders, the regulation became more protective. Instead of being paid the "prevailing wage in the region," which could be as low as $750 a month for about 40 to 60 hours of work per week, the regulation provided that they had to be paid at least the federal minimum wage.[74] This drastic change in the regulation, solely impacting shepherds and herders, provides a unique opportunity to review the impact of increasing wages and bettering regulations on the labor market for H-2A workers.[75]

For farm labor advocates, the 2015 regulatory change—more protective of H-2A workers—would shield US workers from the adverse effects of having to compete with H-2A workers, who accept very low wages. In effect, it would decrease the demand for H-2A workers and increase the demand for US workers.

To assess the impact of the change in the regulation on labor market outcomes, I use a difference-in-differences approach. Essentially, this approach consists of comparing the differences in outcomes for groups who have been targeted by the change in the regulation (herders) with those who were not impacted by the change (other H-2A workers such as crop workers). Specifically, herders were particularly disfavored by the H-2A regulation as they were underpaid and overworked.[76] The regulatory change was meant to correct that.

### A. Treated and Nontreated Groups: Descriptive Statistics

The treated group targeted by the change in the regulation (herders) was particularly disfavored before the change. As Table 4 shows, before the 2015 change, herders earned \$9.15 per hour on average (with some shepherds earning as low as \$4.5 per hour)[77] compared to other H-2A workers who earned an average of close to \$11.2 per hour. In terms of hours worked, herders were working about 44.5 hours a week, about three more hours than other H-2A workers. Herders were in slightly less demand than other H-2A workers (calculated by the number of job vacancies).

**Table 4. Descriptive statistics for treated and untreated groups, before and after the change in regulation (2014-2019) (N= 105,829)**

| | **H-2A Petitions [All periods]** | |
|---|---|---|
| | **Mean (Treated) (N=21,343)** | **Mean (Untreated) (N=84,486)** |
| Hourly wage (in \$) | 10.08 | 11.9 |
| Weekly hours worked | 42.8 | 41.4 |
| Job vacancies (in log) | 0.85 | 1.98 |
| | **H-2A Petitions [Before 2015]** | |
| | **Mean (Treated) (N=21,343)** | **Mean (Untreated) (N=84,486)** |
| Hourly wage (in \$) | 9.15 | 11.2 |
| Weekly hours worked | 44.5 | 41.3 |
| Job vacancies (in log) | 0.93 | 1.32 |
| | **H-2A Petitions [After 2015]** | |
| | **Mean (Treated) (N=21,343)** | **Mean (Untreated) (N=84,486)** |
| Hourly wage (in \$) | 10.13 | 11.9 |
| Weekly hours worked | 44.9 | 44.4 |
| Job vacancies (in log) | 0.84 | 1.98 |

After the 2015 regulatory change, wages were slightly higher for herders and had also increased for other H-2A workers. Because of normal yearly increases in wages, it's difficult to attribute this increase to the regulatory change. However, Table 4 shows that the number of weekly hours worked by herders has increased from 44.5 hours before the change to almost 45 hours a week after the change. The increase is even more striking for nonherder H-2A workers: they work three more hours a week than they did before the change. The demand for herders has decreased while the demand for other H-2A workers has increased.

### B. Empirical Strategy

The baseline specification is as follows:

$$Y_{g,t} = \alpha + \beta\, Treated_g + \delta\, Reg_t + \gamma(Treated_g * Reg_t) + X_{g,t} + \rho_m + \varphi_f + \varepsilon_{g,f,m,t} \quad (1)$$

In this equation, $Y_{g,t}$ is the outcome to be estimated such as the number of job vacancies and hours worked. Job vacancies help determine whether the increased regulation led to a decreased demand for H-2A workers, which is one of the goals of the regulation.[78] The variable $Treated_g$ is a dummy indicating the treated group (herders), which has benefited more from the change in the regulation, notably with wages. $Reg_t$ is a dummy variable for observations coming from the 2015 period. The coefficient of interest, $\gamma$, measures the impact of the change in the regulation on groups of H-2A workers who are herders as opposed to nonherder H-2A workers who did not get benefits from the change in the regulation. $X_{g,t}$ is a set of controls, $\rho_m$ are market fixed effects, $\varphi_f$ are employer fixed effects, and $\varepsilon_{i,f,m,t}$ is the error term.

A few endogeneity concerns arise from this difference-in-differences specification. Individuals might have specific characteristics such as training, more years of education, health condition, and age that might affect and determine their wage income. Therefore, I controlled for those individual characteristics. The set of control variables also includes dummies for whether the herder was hired through an association or directly by a farmer. I also controlled for employer-specific characteristics and included employer and market fixed effects. These controls account for the possible endogeneity in the model. The results are robust to the addition of all these controls.

## C. Results

Table 5 shows the results of the difference-in-difference estimates from Equation (2). More precisely, the coefficient of interest—the difference-in-difference estimator—is Treated*Reg and is defined as the difference in average outcome in the treatment group (herders) before and after the regulatory change minus the difference in average outcome in the nontreated group (other, nonherder H-2A workers) before and after the regulatory change.

The results in column (1) show that the new regulation had a negative and significant impact on the total job vacancies within a market. Thus, as predicted, the increase in wages and more protective law led to a very sharp decrease (about 75%) in job vacancies (which is our proxy for demand for H-2A herders) compared to the nontreated group. It's possible that farmers turn instead to US workers to do these jobs or ask their herders to work more hours. In fact, column (2) shows that, after the 2015 regulatory change, herders' hours have considerably and significantly increased (about 27%) compared to other nonherder H-2A workers.

**Table 5. Difference-in-differences estimates**

| **Dependent Variables** | **(1) Log job vacancies** | **(2) Log hours** |
|---|---|---|
| Treated | –0.3911*** | –0.0053 |
| | (0.049422) | (0.00463) |
| **Treated*Reg** | –0.7500*** | 0.2716*** |
| | (0.050206) | (0.004728) |
| Additional Controls | Y | Y |
| Firm Fixed Effects | Y | Y |
| Market Fixed Effects | Y | Y |
| Adj R-squared | 0.12 | 0.11 |
| N | 105,829 | 105,829 |

Notes: Robust standard errors are in parentheses and clustered at the county level.
***, **, and * indicate significance at the 1%, 5%, and 10% confidence levels.

This sharp decrease in demand for herders is particularly striking and may suggest a few things. First, it is possible that employers do not really need herders as much as they lead the DOL to believe during the labor certification process. If they are able to do away with these workers once wages increase and working conditions improve, then it is possible that employers did not need to recruit as many herders to begin with.

Second, it is possible that employers switched from employing H-2A herders to employing US workers. If this is the case, then the regulation has had its intended effect—to protect the US workforce. However, if employers are able to swiftly adapt and recruit US workers, it highlights the fact that the labor certification process is riddled with fraud and that there is no labor shortage. Another alternative is that employers are hiring undocumented workers instead. Because there are very little data on undocumented workers [besides data recording arrests from the US Immigration and Customs Enforcement (ICE)], it is difficult to determine whether the decrease in demand for H-2A herders led to an increase in demand for undocumented workers.

Third, employers might employ fewer but more productive herders. For instance, herders, usually from Peru, have some experience herding in Peruvian mountains. This experience may be very valuable to ranchers in Colorado or Wyoming, who often rely on H-2A herders. Instead of employing many H-2A herders (some with experience, some without experience), it's possible that employers are hiring fewer H-2A herders but more expert herders. While I controlled for some individual characteristics, the data do not provide information on specific herding-related skills any of the workers may have, and that could be influencing those estimates.

## VII. H-2A Jobs Are Primarily in Concentrated Labor Markets

Although there is some debate on market definition and on whether labor markets are more concentrated in general, there is substantial evidence that labor market concentration is negatively correlated with wages.[79] Generally, specialized labor markets are more prone to employer power, and perhaps even more so for low-skill workers.[80] Often, it is linked to lower wages and poorer working conditions. As employers have more power (because they hire a large percentage of workers in a market), they tend to worsen working conditions. The lack of competition between firms leads to the overall degradation of wages and labor conditions.[81]

Few studies have looked at the effects of labor market power on guestworker programs in the United States. Some scholarship has assessed the impact of monopsony power on undocumented immigrants and has shown that undocumented workers are more vulnerable to firms' market power.[82] In Germany, researchers found that undocumented workers were likelier to face monopsonistic discrimination than their native counterparts.[83] Similarly, in the United Arab Emirates, low-skilled guestworkers are more vulnerable to firm power than other workers. Further, some of the frictions from these guestworker programs contribute to lowering wages.[84] A recent paper analyzes the effects of labor market monopsony on H-1B, H-2B, and H-2A workers,[85] and the authors find that guestworkers face moderately or highly concentrated labor markets and that it decreases wages. However, the bulk of their data mostly deal with H-1B and H-2B visa programs. H-2 A programs vary considerably from other guestworker programs. Not only are H-2A visas granted for very short periods (about nine months usually), but H-2A workers cannot switch jobs or move to another part of the country while remaining legally in the United States.

The history of the H-2A program as well as the *Llacua* case strongly suggest the presence of a monopsonistic labor market in which farmers may enjoy some degree of wage-setting power. As discussed later, collusion and fraud have led to a more concentrated labor market in regions that hire H-2A workers. In this context, this paper attempts to assess labor market concentration in the agriculture sector for specific occupations that import H-2A workers and whether such concentration has effects on H-2A workers.

## *A. Descriptive Statistics*

**Table 6. Summary statistics (N=43,643)**

| **H-2A Petitions** | | |
|---|---|---|
| | Mean | SD |
| Hourly wage (in $) | 11.86 | 1.15 |
| Concentration (HHI) | 0.25 | 0.28 |
| Weekly hours worked | 41.6 | 6.5 |
| **Categorical Variable – H-2A Workers** | | |
| **Education** | | |
| None | 98.8% | |
| High School | 1.17% | |
| Bachelor | 0.02% | |
| **Occupation and SOC code** | | |
| Farmworkers and Laborers, Crop, Nursery, and Greenhouse (45-2092) | 64% | |
| Farmworkers, Farm, Ranch, and Aquacultural Animals (45-2093) | 20% | |
| Agricultural Equipment Operator (45-2091) | 14% | |
| Other (various but agriculture related) | 2% | |
| **Categorical Variable – H-2A Petitioner** | | |
| **Type of Petitioner**[86] | | |
| Individual Employer | 69% | |
| Association – Filing as Agent | 15.33% | |
| H-2A Labor/Job Contractor | 15% | |
| Association – Sole Employer | 0.42% | |
| Association – Joint Employer | 0.11% | |
| **Training Requirement (none)** | 99% | |

The outcome measure, hourly wage, is disclosed by the employer. The average hourly wage in the sample is $11.86 per hour, and workers work an average of 41.6 hours a week. As discussed, and as it was exemplified in the *Llacua* case, associations of farmers and ranchers can often file on behalf of their members and act as employees. The dataset provides information about whether the application was filed by an association, by an individual employer, or by an H-2A contractor.[87] In the sample, the vast majority of employers file individually, that is, not through an association. About 16% of employers file applications through an association, and about 15% use contractors.

Additional control variables, such as the log of county total population and the log of the commuting zone population are taken directly from the Bureau of Labor Statistics and merged with the data. Other individual controls such as education level, training, and weekly hours worked come from the data.

In terms of occupations, all H-2A workers work in the agriculture sector, but they have different jobs. A large majority of H-2A workers in the dataset are farmworkers and laborers (84%). Among those workers, approximately 64% plant, cultivate, and harvest crops, vegetables, fruits, and nuts (these workers are classified as Farmworkers and

Laborers, Crop, Nursery, and Greenhouse (45-2092) in the sample). Some of their daily tasks may include shoveling, tilling soil, applying fertilizer, weeding, thinning, or pruning crops. They may also harvest, load, and package harvested products. About 20% of the farmworkers and laborers care for animals and livestock (these workers are classified under Farmworkers, Farm, Ranch, and Aquacultural Animals (45-2093) in the sample). They attend to live farming, ranching, open range, or aquacultural animals such as sheep, goats, or bees. Some of their duties can include feeding, watering, herding, and grazing the animals. They may also examine animals to detect diseases and injuries and assist in births. About 14% of workers are tasked with operating agricultural equipment (these workers are classified under Agricultural Equipment Operator (45-2091) in the sample). It includes planting, cultivating, feeding, and herding livestock; crop baling; and hay bucking. Almost no job requires past training or a specific education level.

### B. The Labor Market for H-2A Jobs Is Concentrated

The labor market for jobs that often require H-2A workers is moderately to highly concentrated. To measure concentration within a specific labor market, I use the Herfindahl-Hirschman Index (HHI), calculated based on the share of vacancies of all employers that post vacancies in that market.[88] Job vacancies are a good proxy for determining whether an employer has a large market share in a specific labor market. If, for example, an employer is responsible for hiring 80% of shepherds in a geographic area, then it is likely that it has most of the market share.

Theoretically, the HHI is directly linked to wages in the Cournot model of oligopsonistic competition.[89] An increase in HHI leads to an increase in the gap between marginal productivity of labor and wages, according to Boal and Ransom.[90] In the literature on labor market monopsony, the HHI is a measure of labor market concentration that is widely used.[91]

Following Azar et al.,[92] I chose to define work (or occupation) as a six-digit Standard Occupation Classification (SOC)[93] as the baseline. Although the authors' calculations show that a better and more precise measure are job titles, they explain that using the six-digit SOC is a conservative choice. Because it is too broad, the concentration measured by the HHI will probably be underestimated. Gibbons et al. also follow such an analysis.[94] In any event, the LCA data do not easily record job titles. As a preliminary analysis, it is therefore satisfactory to use the six-digit SOC.

Thus, formally, the HHI is defined as follows:

$$HHI_{m,t} = \sum_{j=1}^{J} s_{j,m,t}^{2}$$

where for any firm *f* in a given six-digit SOC market *m* and at year *t*, its market share of job postings for the six-digit SOC market is defined as the number of vacancies or the number of job postings by the firm in the market in year *t* divided by the number of all job postings or vacancies in that six-digit SOC market in year *t*.

HHI scores are between zero and one and increase with the level of concentration in a market. The Department of Justice (DOJ) and Federal Trade Commission's (FTC) Horizontal Merger Guidelines[95] consider that a market is moderately concentrated if the HHI is above 0.15, and highly concentrated if the HHI is above 0.25. In the sample, on average, the HHI is equal to 0.25, which means that the market is moderately concentrated but right at the limit of high market concentration. Figures 1, 2, and 3 in the Appendix show histograms of labor market concentration for the three main SOCs in the data; namely, farmworkers in farms and ranches (45-2093), agricultural equipment operators (45-2091), and farmworkers in crops and nursery or greenhouses (45-2092). Figure 1 shows that the labor market for equipment operators is moderately concentrated, with the distribution of the HHI being either below 0.25 or at 0.5 or close to 1. Figure 2 shows that the labor market for farmworkers harvesting crops or working in greenhouses is moderately concentrated, with an average HHI of 0.17. Such farmworkers represent 69% of the sample, so the HHI measured here is reliable. Finally, Figure 3 shows that the market for farmworkers in farms or ranches is the most concentrated, with an average HHI of 0.38.

In this framework, there is virtually no worker mobility. Unlike most other workers, H-2A workers cannot switch jobs. They are constrained to work for a particular employer in a specific geographic area and for a specific length of time. This makes the analysis a little easier since I don't have to worry about the fact that monopsony and market power could make the observed transition rates endogenous (there are no transition rates because workers can't switch jobs).

## *C. Labor Market Concentration Is Negatively Associated with H-2A Workers' Wages*

Based on the historical and legal framework of H-2A programs and on some preliminary evidence of labor market concentration in the agriculture sector, specifically for occupations using H-2A workers. To test the validity of this evidence further, I conducted regressions that controled for commuting zone by occupation effects and attempted to isolate the effect solely from variation in HHI and wages over time within a given commuting zone and occupation. The baseline equation, similar to Azar et al.,[96] is as follows:

$$Log(w_{i,f,m,t}) = \alpha + \beta logHHI_{m,t} + \gamma X_{i,f,m,t} + \delta_t + \rho_m + \varphi_f + \varepsilon_{i,f,m,t}, \quad (1)$$

where $Log(w_{i,m,t})$ is the log real wage for H-2A job a petitioner/worker $i$ in market $m$ in year $t$, $logHHI_{m,t}$ is the corresponding $logHHI_{m,t}$ within the market $m$ in year $t$, $X_{i,m,t}$, is a set of controls, and $\delta_t, \rho_m, \varphi_f$are year, market, and firm-specific fixed effects. $\varepsilon_{i,f,m,t}$ is the error term.

I first ran a specification by dividing the sample for the three most represented occupations in the sample (SOC 45-2091, 45-2092, and 45-2093) with all fixed effects. Figures 1, 2, and 3 show more concentrated markets for occupation 45-2093, so I anticipated that the largest effect would be on the wages of such workers. I clustered standard errors at the commuting zone/occupation level. The second specification I ran focused on subsamples of H-2A petitions characterized by how they were filed: by an individual employer, by an association, or by a contractor.

## *D. Results*

One of the main results in this section is that higher levels of labor market concentration are associated with lower wages. Results are significant for the overall sample. Table 7 shows the results for all occupations in column (1), in which for every percentage of increase in labor market concentration leads to a very small decrease in wages. In column (2), the association between an increased HHI and wages is also negative, but it becomes insignificant when adding market fixed effects and firm fixed effects, which means that the negative relationship was probably driven by variation in H-2A petition wages. Column (3) shows that moderate concentration in the market leads to a decrease in wages among harvesters and farmworkers that is significant although very small in magnitude. Last, column (4) shows that the most concentrated market (that of farmworkers in ranches, such as sheepherders in the *Llacua* case) has a negative effect on wages even if, again, the magnitude is relatively small.

**Table 7. Descriptive regressions of labor wages on labor market concentration**

| Dependent Variable: Log hourly wage | (1) All workers | (2) Agriculture operators (45–2091) | (3) Crops (45–2092) | (4) Herders (45–2093) |
|---|---|---|---|---|
| Log Labor Market HHI | −0.00307*** | −0.00014 | −0.00480*** | −0.017*** |
| | (0.000499) | (0.00166) | (0.00087) | (0.00446) |
| Additional Controls | Y | Y | Y | Y |
| Firm Fixed Effects | Y | Y | Y | Y |
| Market Fixed Effects | Y | Y | Y | Y |
| Year Fixed Effects | Y | Y | Y | Y |
| Adj R-squared | 0.856 | 0.907 | 0.902 | 0.707 |
| N | 43,304 | 5,249 | 28,447 | 8,767 |

Notes: The table presents estimates of the hourly wage with respect to local labor market concentration and variation in this estimate across places with differing labor market concentrations for SOC 45-2091, 45-2092, 45-2093.

Table 8 columns (2), (3), (4) focus on a subsample of H-2A petitions filed by individual employers, contractors, or associations, respectively. For H-2A petitions filed by individual employers (1) or by a contractor (3), market concentration did not have a significant effect on wages. However, in the subsample of workers who filed through an association, an increase in one log point of HHI is associated with a small decrease in wages that is significant and consistent with the results in Table 2. Furthermore, those results are of a similar magnitude as the results in previous studies using a similar methodology.[97]

The first results of this study show that increased levels of labor market concentration in the agriculture sector are associated with decreased wages for H-2A workers. This result is in line with the literature on the effects of concentration on wages,[98] and the latest study on labor market monopsony and guestworkers in the United States.[99] In fact, in the H-2A program context, wage suppression has been occurring since the early days of the program. An interesting new result in this paper is the role that the hiring associations might be playing and how a new definition of market power might be in order when analyzing the H-2A program. This new definition might enable courts to analyze labor market antitrust cases differently and reconcile immigration law and antitrust law.

**Table 8. Descriptive regressions of labor wages on labor market concentration for subsamples of filing organization**

| Dependent Variable: Log hourly wage | (1) All workers | (2) Individual Filed | (3) Contractor Filed | (4) Association Filed |
|---|---|---|---|---|
| Log Labor Market HHI | −0.00307*** | 0.0011 | 0.00760 | −0.0181** |
| | (0.000499) | (0.00416) | (0.00620) | (0.00597) |
| Additional Controls | Y | Y | Y | Y |
| Firm Fixed Effects | Y | Y | Y | Y |
| Market Fixed Effects | Y | Y | Y | Y |
| Year Fixed Effects | Y | Y | Y | Y |
| Adj R-squared | 0.856 | 0.865 | 0.427 | 0.995 |
| N | 43,304 | 22,157 | 5,112 | 7,827 |

Notes: The table presents estimates of the hourly wage with respect to local labor market concentration and variation in this estimate, depending on whether an individual employee filed the application (Column 2), an employer went through a contractor to hire H-2A workers (Column 3), or an employer went through an association to hire H-2A workers (Column 4).

Defining market power is one of the most important pieces of the puzzle. Using the proper definition allows us to determine whether the employer exerts dominance over a specific market. Courts have determined that a definition of the relevant market is necessary when analyzing antitrust cases. With H-2A workers, defining the relevant market power might mean defining the associations of ranchers that file applications to recruit H-2A workers. Showing that those farmers represent a large market share, as in the *Llacua* case, would help determine market power. Redefining market power to understand associations' roles might be a way to reconcile antitrust law and immigration law. Even if the immigration statute allows associations to act as employers, it does not prevent courts from analyzing the market structure to "determine whether it is "susceptible to the exercise of market power through tacit coordination." To do so, a court has to show that the "[s]usceptible markets tend to be highly concentrated—that is, oligopolistic—and to have fungible products subject to inelastic demand."[100]

Economists and legal scholars have determined that "a labor market is concentrated when a few firms dominate hiring in the market."[101] In the *Llacua* case, the two rancher associations—the WRA and the MPAS—are responsible for hiring 91% of herders. That is a very concentrated market. Almost no herder will find work within this region without going through either the WRA or the MPAS.

H-2A jobs are probably highly fungible, which makes it easier for collusion to occur. The job titles and job descriptions, from rancher to rancher, are the same. Further, courts have suggested that even if "[t]he jobs in question may not be inherently fungible," the purpose "is to test whether defendants would be able to compare the positions for coordination purposes." In *Llacua,* for example, ranchers go through associations to recruit workers, and associations then allocate workers to ranchers in a nonindividualized way, so even if a particular worker is not the same as another one, the work product is the same and in that sense, workers are fungible.

Following this reasoning, associations are allowed to act as employers and are therefore supposed to be treated as such when defining labor market power. Market definition might need to be changed to account for these "employers" who may be employing a large portion of H-2A workers in certain markets. If this is the case, then the estimates in Tables 2 and 3 are conservative. Only a few associations operate in a given market. Grouping member-employers together by association might lead to results of a higher magnitude since labor markets will likely be more concentrated.

## VIII. Further Discussion and Limitations

This section discusses some of the limitations or possible alternative explanations for the empirical results explained in this paper. For instance, employers' reliance on undocumented workers and the existence of network effects could potentially explain some of the results. Additionally, it is important to explain data limitations and how these limitations might influence the results. Last, this section will discuss some of the policy implications of the results and some areas for further research.

### *A. Undocumented Workers*

As discussed, one of the H-2A program's goals was to reduce the entry of undocumented noncitizens. In the past, some critics have argued that improving the regulation and increasing wages for H-2A workers would make the program prohibitively expensive for employers.[102] It is possible that strengthening the regulation will have perverse effects.

For instance, if wages are higher, employers might increasingly rely on undocumented workers instead of hiring US workers or paying their workers proper wages. In the difference-in-differences estimates, it is therefore possible that the decrease in job vacancies for herders after the 2015 regulatory change did not lead to increased hiring of US workers. Instead, the demand for workers could have shifted to undocumented workers. In other words, it is possible that in order to keep wages and working conditions low, employers hire more undocumented workers. While this would put them at some legal risk, given the overall lack of enforcement, the risk is probably low for these employers.

And, although the number of undocumented workers has decreased over the past years, undocumented workers still held about a quarter of farming jobs in 2016.[103]

If this is the case, then the 2015 regulation, or any regulation that attempts to affect labor policy and H-2A workers, might not have the intended effects since employers can hire outside of the legal labor market. Perhaps what is deeply needed is not more regulation but enforcement of the existing regulation and close monitoring of any fraud in the labor certification process.[104]

### B. Network Effects

It is possible that the concentration of H-2A workers and the reliance on H-2A workers is due to the presence of network effects. If H-2A workers know that the geographic area they will be working in might have other workers from their home country, they might be more inclined to work in those specific areas. Although such effects have probably not biased the results in this paper since market fixed effects were included in all estimations, it is an important element to highlight, especially when discussing migration and migrant labor.

In fact, studies have shown that H-2A workers do not necessarily favor "states where they can leave work and readily blend in" or that specific existing migrant networks attract H-2A workers.[105] These findings are unsurprising. H-2A workers are needed in areas where there is a high demand for agricultural work. For instance, states that certify most H-2A positions rely heavily on agriculture. The top five states that import H-2A workers are Florida, Georgia, Washington, California, and North Carolina.[106] It is likelier that those workers end up in those geographic areas because they are needed rather than because there are network effects at play there (although it's possible that the presence of H-2A workers leads to network effects in the long run).

### C. Limitations

#### *1. Data Limitations*

The data I have are from USCIS, which computes most information from H-2A visa petitions. However, in the second section of this paper, I described how H-2A workers are subject to wage theft, fraud, and coercion and can sometimes be paid as low as $1.50 per hour (after illegal kickbacks). The USCIS data do not account for illegal practices. If anything, my results paint a realistic picture for H-2A workers who work for law-abiding employers, but do not account for any illegal practices employers might engage in. The lack of enforcement of the regulation might taint what the results describe.[107]

Further, the data provide a partial view of the dynamics at play. Because the data exclusively focus on H-2A workers, they fail to account for general labor market effects that include US workers. This is a limitation that I was aware of from the start and tried to correct by using other data. However, most publicly available data are aggregated, which makes individual-level analysis impossible. Individual-level data for US workers are difficult to gather and require an extensive application.[108] For now, the analysis presents only a partial view of what might be going on. For instance, the difference-in-differences estimates show a decline in the number of job vacancies for H-2A workers. From the estimates, it's unclear whether the regulation has instead increased job vacancies for US workers. Without knowing more on how the regulation impacts US workers, it is difficult to explain whether the strengthening of the 2015 regulatory change made US workers better off.

#### *2. The Parallel Trend Assumption*

To be convincing, the difference-in-difference estimates in Equation (1) have to satisfy the key identifying assumption that the outcome trends would have been the same among individuals in the treated and control groups in the absence of treatment (i.e., the parallel trend assumption). In this paper, because I have only one wave of data before the regulatory change, the parallel trend assumption is difficult to verify. To make sure that the estimates are not biased, I have included variables that control for individual time-varying characteristics, industry fixed effects, and county fixed effects [see Equation (2)]. More data are available before 2014, and a future version of this paper will be able to show whether the parallel trend assumption is satisfied.[109]

*3. Robustness Checks*

To make sure that the results are robust, it was useful to run a couple of sensitivity tests that are quite common when using difference-in-difference approaches.[110] For example, using a different control group or a placebo treatment group can help determine whether estimates are driven by other trends in the data not related to the regulatory change. While further versions of this paper will hopefully include additional waves of data to show compliance with the parallel trend assumption, the robustness checks described here are also a helpful way to alleviate any concerns regarding biased estimates.

## IX. Policy Considerations

### *A. Benefits of Remittances*

Despite the widespread abuses taking place within the H-2A program, guestworker programs (such as the H-2A program) bring some benefits to those workers. Growers and ranchers have noted that "their foreign workers routinely send funds home, suggesting that the herders have expendable income."[111] For H-2A workers who are not subject to wage theft and other abuse, their earnings might be a valuable source of income for their family members in their home country.

There is vast literature on the benefits of remittances. Some studies have found that remittances relax the household income constraints and allow children more educational attainment.[112] Moreover, some scholars have shown that temporary worker programs can have positive effects on "income, consumption, durable goods consumption, and subjective standards of living".[113] Other studies argue that, although temporary worker programs may provide guestworkers with money to remit home, they "do little to increase actual economic development in the sending communities, and they are not truly temporary in nature."[114] In fact, the studies highlight that those programs actually "perpetuate migration, both legal and illegal, by inducing a reliance on income that can be earned only abroad."[115]

While this paper's goal is not to focus on the benefits of the H-2A program on workers' home countries, any policy decision should account for the benefits or loss to those communities.

### *B. Other Welfare Effects*

The H-2A program also has other benefits and drawbacks as it influences consumers, employers, domestic workers, and foreign workers in the economy. In this subsection, I describe some of the welfare effects the H-2A program has nationally. A policymaker should keep those welfare dynamics in mind when changing the regulation.

The first and relatively straightforward benefit is that US consumers benefit from lower prices due to the lower costs of labor. By importing a foreign workforce, firms are able to considerably reduce their production costs because they pay lower wages than they would have to pay if they were trying to recruit US workers. Presumably, if employers are able to lower the costs of production, it also drives down the price of goods, which increases consumer surplus (or consumer welfare).

On the other hand, domestic workers who compete with foreign workers for the same jobs are harmed in two ways. Because there is an increased supply of workers for a given occupation, the wage for that occupation is driven down. So, in a given labor market, the increase in labor supply will shift the supply curve to the right, and wages will fall. Also, given that an increase in workers will decrease the price of goods, there is an ambiguous effect on employment and hours worked for domestic workers. For example, consider a decrease in the price of goods.[116] Such a decrease will increase the domestic worker's reservation wage. In fact, the worker will be able to buy more goods for the same amount of work. It will make the worker less likely to work because she can satisfy the same level of utility with less work. To be willing or motivated to work, the reservation wage will therefore have to be higher (the worker will be willing to forgo leisure to work). In turn, this will lead to a decrease in labor supply or a decrease in domestic employment. Further, the price decrease increases domestic workers' real wages (they can buy more with the same wage), which decreases the demand for leisure and leads to more hours worked.[117]

Last, domestic workers benefit from the increased demand for their labor. This scenario happens *only* when foreign workers are complements to domestic workers.[118] An influx of foreign workers will positively impact domestic workers who are complements to the incoming foreign workers, as the demand for domestic workers will also likely increase. Additionally, if foreign workers consume goods produced by domestic workers, then it is possible that the demand for domestic workers for the production of these goods will also increase. This would mitigate the negative effects of an influx of foreign workers.

## X. Conclusion

This paper attempted to give a descriptive empirical account of the law-and-economics dynamics behind the H-2A program. It also contributes to the literature on guestworker programs. By measuring the impact of a significant regulatory change in the program, this paper finds that protective regulatory changes might have some of their intended effects and lead to less labor demand for foreign workers. However, it also leads to significantly more hours worked for those workers. Additionally, this paper finds that the fraudulent and collusive behavior that occurs throughout the labor certification process might lead to increased labor market concentration in markets that recruit H-2A workers. In turn, increased labor market concentration is associated with small decreases in wages for H-2A workers.

These results explain only part of the situation: the situation of H-2A workers. Future research should attempt to look at whether decreased labor demand for H-2A workers when regulations become more protective of workers contributes to increased demand for US workers. Further, future research should analyze the effects of labor market concentration in the agricultural sector on US workers and the dynamics between H-2A and US workers.

Finally, any changes in the H-2A regulation should account for possible welfare effects. The regulation, as is, has ambiguous welfare implications: it benefits consumers, employers, and perhaps H-2A workers (if they are paid correctly). The welfare effects on US workers—who are also consumers—are necessarily negative unless US workers are complements to H-2A workers. The program should be updated to account for economic dynamics and their potential consequences.

## Acknowledgments

Many thanks to Professor Ginsburg, Professor Kim, Dean Miles, and Professor Fiona Scott-Morton for their comments and feedback. Thanks also to participants at the Yale Big Ag & Antitrust Conference and to my wonderful classmates in Canonical Ideas for their insightful comments.

## Appendix

Figure 1. HHI levels among 45-2091 workers (mean=0.22)

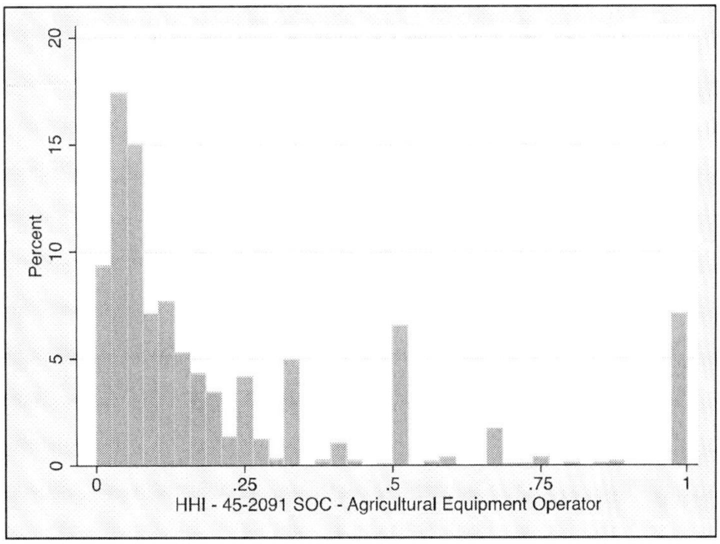


Figure 2. HHI levels among 45-2092 workers (mean=0.17)

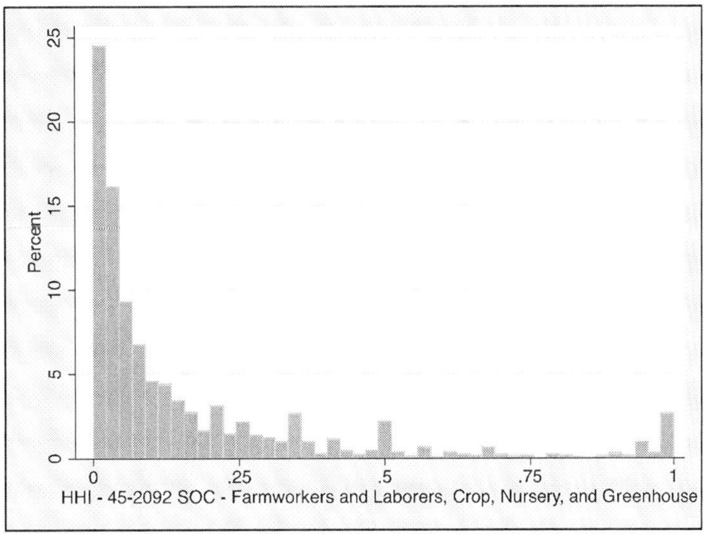

**Figure 3. HHI levels among 45-2093 workers (mean=0.38)**

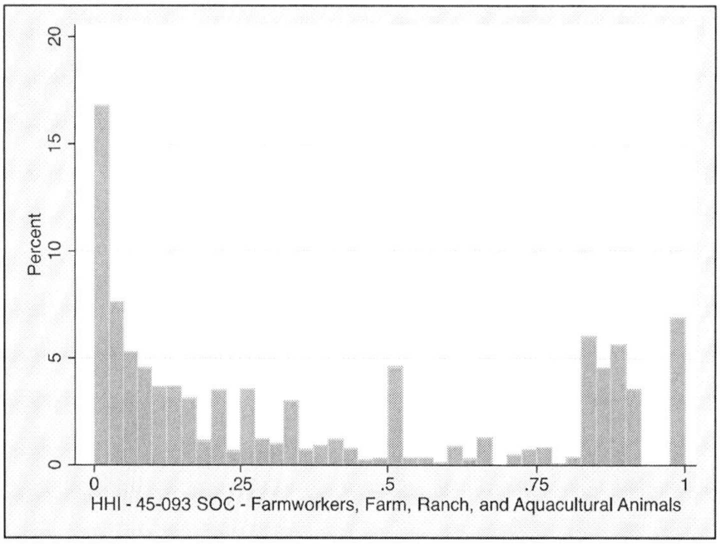


**Figure 4: H-2A Visas Issued, FY 1992-FY 2019**

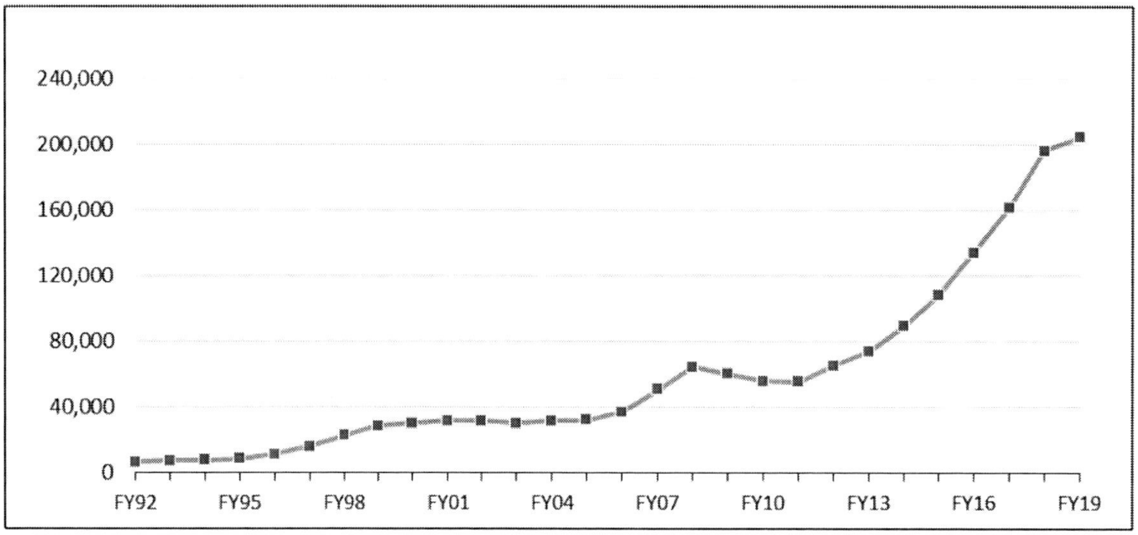


Source: CRS presentation of data from US Department of State, Bureau of Consular Affairs.

**Figure 5. Administrative Process to Bring in H-2A Workers**

Source: CRS presentation of information from US Department of Homeland Security, USCIS.

## Endnotes

[1] Labor Certification Process for Temporary Agricultural Employment in the United States (H-2A Workers), 20 C.F.R. § 655.10 (2015).
[2] *See, e.g.*, Glen Weyl, *The Openness-Equality Trade-off in Global Redistribution*, 128 ECON. J. 1 (2018); Edward Kosack, *Guest Worker Programs and Human Capital Investment: The Bracero Program in Mexico, 1942–1964*, 56 J. Human Resources 570, 574 (2020).
[3] *See generally* Centro de los Derechos del Migrante, *Ripe for Reform* (2020), https://perma.cc/8UVN-HGR8.
[4] *See generally See* Alison K. Guernsey, *Double Denial: How Both the DOL and Organized Labor Fail Domestic Agricultural Workers in the Face of H-2A*, 93 Iowa L. Rev. 277, 292–95 (2007).
[5] *See* Alessandra Casella & Adam B. Cox, *A Property Rights Approach to Temporary Work Visas*, 47 J. L. Studies 1, S201–202 (2018).
[6] *See* Eric M. Gibbons, Allie Greenman, Peter Norlander and Todd Sørensen, *Monopsony Power and Guestworker Programs*, 64 The Antitrust Bulletin 540 (2019).
[7] Pierre Cahuc, Stéphane Carcillo & André Zylberberg, Labor Economics 159 (2nd ed.) (2014).
[8] *See The Bracero Program*, UCLA Labor Center, https://perma.cc/47NY-B2K2; Garry G. Geffert, *H-2A Guestworker Program, A Legacy of Importing Agricultural Labor*, *in* The Human Cost of Food 113 (Charles D. Thompson, Jr. & Melinda F. Wiggins, eds., 2002).
[9] *See* Table 1, *infra.*
[10] 8 U.S.C. § 1182(a)(14); § 1188(a).
[11] Andorra Bruno, *H-2A and H-2B Temporary Worker Visas: Policy and Related Issues*, Congressional Research Service 7 (2020).
[12] Andorra Bruno, *Immigration of Temporary Lower-Skilled Workers: Current Policy and Related Issues*, Congressional Research Service 3–4 (2012).
[13] *See id.* at 4.
[14] The steps are visually represented in Figure 5 in the appendix, *infra.*
[15] Please note, though, that there is a difference between the number of petitions certified and the number of visas issues in the end (see Figure 4 in the Appendix). A worker may end up not getting their visa approved for a variety of reasons independent from the labor certification process.
[16] *See* USCIS, *Immigration and Citizenship Data*, https://perma.cc/UNA8-E8R3.
[17] *See, e.g.*, Digilab, Inc. v. Sec'y of Lab., 495 F.2d 323 (1st Cir. 1974).
[18] 419 U.S. 65 (1974).
[19] *See id.* at 76.
[20] *Silva v. Secretary of Labor*, 518 F2d. 301 (CA1 Mass. 1975).
[21] 8 U.S.C. § 1182(a)(14).
[22] Matter of Gonzalez-Becerra, 13 I. & N. Dec. 387, 387 (BIA 1969).
[23] § 1188 provides the same language for labor certification as § 1182(a)(14).
[24] 567 F.2d 1154 (1st Cir. 1977).
[25] *Hernandez Flecha*, 567 F.2d at 1156.
[26] 501 F.2d 757 (D.C. Cir. 1974).
[27] 508 F.2d 493 (1st Cir. 1974).
[28] *Elton Orchards, Inc.*, 508 F.2d at 500.
[29] *Rogers*, 563 F.2d at 626; *see also supra* Part I.A.
[30] *Rogers*, 563 F.2d at 626.
[31] *See* Cahuc, *supra* note 7.
[32] *Id.*

[33] *See* Alessandra Casella & Adam B. Cox, *A Property Rights Approach to Temporary Work Visas*, 47 J. L. Studies 1, S201–202 (2018).
[34] *See* Cahuc, *supra* note 7.
[35] 8 U.S.C. § 1188(a)(1)(A).
[36] A SWA is a state government agency that receives funding to administer the state's public labor activities pursuant to the Wagner-Peyser Act (29 U.S.C. § 49 et seq.).
[37] CalJobs (search terms used: "harvester" and "farming, fishing, and forestry occupations" category), https://perma.cc/Q4X4-YHU7.
[38] *Id.* (search terms used: "herder" and "farming, fishing, and forestry occupations" category), https://perma.cc/J8YD-Z6MY.
[39] *See* Connecting Colorado, https://perma.cc/U6M7-JE6D.
[40] *See* Lauren A. Apgar, *Authorized Status, Limited Returns*, Economic Policy Institute (May 21, 2015), https://perma.cc/B5XW-FYQG.
[41] *See* Alison K. Guernsey, *Double Denial: How Both the DOL and Organized Labor Fail Domestic Agricultural Workers in the Face of H-2A*, 93 Iowa L. Rev. 277, 292–95 (2007).
[42] *See* 8 U.S.C. § 1188(d).
[43] *See id.* If an association is a joint or sole employer of temporary agricultural workers, the certifications granted under this section to the association may be used for the certified job opportunities of any of its producer members and such workers may be transferred among its producer members to perform agricultural services of a temporary or seasonal nature for which the certifications were granted.
[44] 930 F.3d 1161 (10th Cir. 2019).
[45] *See Llacua*, 930 F.3d 1161 (holding that the associations did not violate Section 1 of the Sherman Act). *Llacua* is the only case that challenged agricultural associations, though.
[46] Deborah Cohen, Braceros: Migrant Citizens And Transnational Subjects in the Postwar United States And Mexico 51 (2011).
[47] 125 Cong. Rec. 16789, 16862 (1979) (statement of Sen. Harrison Schmitt).
[48] *See* Jennifer J. Lee, *U.S. Workers Need Not Apply: Challenging Low-Wage Guest Worker Programs*, 28 Stan. L. & Pol. Rev. 1, 26 (2017).
[49] *Id.* at 28.
[50] *See* Lee, *supra* note 48 (quoting Charles Collins Teague, Fifty Years A Rancher 141–42 (1928)).
[51] *See* Frank H. Wu, *Neither Black nor White: Asian Americans and Affirmative Action*, 15 B.C. Third World L.J. 225, 232 (1995).
[52] *See* Lee *supra* note 48 at 36.
[53] *See* Lee *supra* note 48 at 36.
[54] *Id.* at 37.
[55] Marta María Maldonado, Racial Triangulation of Latino/a Workers by Agricultural Employers, 26 HUM. ORG. 353, 358 (2006).
[56] See Lee, supra note 48 at 40.
[57] See generally Centro de los Derechos del Migrante, Ripe for Reform (2020), https://perma.cc/8UVN-HGR8.
[58] See id. at 21.
[59] Id.
[60] Discussing and analyzing these violations in more detail is outside the scope of this paper but it is important to note that the violations contribute to a climate of fear and abuse that prevent H-2A workers from receiving proper pay or claiming wage fraud. More importantly, it perpetuates the exploitation of guestworkers for the benefits of growers and at the expense of US and migrant workers (though in different ways). See id. at 24–26 (reporting that "[t]hirty-one percent of those interviewed [for the report] said they were subjected to serious verbal abuse, including the use of racial epithets.").
[61] See Department of Labor, 2015 Final Rule: Range Herding or Production of Livestock in the United States, https://perma.cc/6PFE-CP65 (hereinafter 2015 Final Rule).
[62] See Complaint at 6, *Llacua v. Western Range Ass'n*, No. 15-cv-01889, 930 F.3d 1161, 1177 (10th Cir. 2019) (hereinafter Llacua Complaint).
[63] US Department of Labor, Temporary Agricultural Employment of H-2A Foreign Workers in the Herding or Production of Livestock on the Range in the United States, 80 Fed. Reg. 62958 (Oct. 16, 2015).
[64] *See* 20 C.F.R. §655.120, §655.12(l). The AEWR is a specific minimum wage for H-2A workers.
[65] *See* 2015 Final Rule, *supra* note 61.
[66] *Performance Data*, US Department of Labor, https://perma.cc/G3YQ-NENY.
[67] Regulation during the COVID-19 pandemic allowed employees to prolong their workers' H-2A visas, for example.
[68] *See* Eric M. Gibbons, Allie Greenman, Peter Norlander and Todd Sørensen, *Monopsony Power and Guestworker Programs*, 64 The Antitrust Bulletin 540 (2019).
[69] *See* S.J. Davis, R.J. Faberman & J.C. Haltiwanger, 2013. The Establishment-Level Behavior of Vacancies and Hiring, 2 Q. J. Econ. 128 (2013); *see also* José A. Azar, Ioana Marinescu, Marshall I. Steinbaum, and Bledi Taska, *Concentration in US Labor Markets: Evidence from Online Vacancy Data*, 66 Labour Econ. (2020).
[70] *See infra* Part V.A.
[71] Such applications represent about 0.5% of the sample size.
[72] This part is meant to introduce the data, which is why it is on the shorter side. The fifth and sixth sections of this paper discuss the data in more detail as well as the particularities of the sub-samples used in the analyses.
[73] *See* 2015 Final Rule, *supra* note 61.

[74] *See Llacua* Complaint, *supra* note 62.
[75] My goal is to also look at the impact on US workers, but I haven't completely cleaned out the data yet (the dataset is really rough).
[76] *See Llacua* Complaint, *supra* note 62.
[77] *See Llacua* Complaint, *supra* note 74 at 10. If there is some discrepancy between the hours worked described in the Complaint in *Llacua* and the descriptive statistics here, it could be due to the fact that this is a dataset from USCIS. Employers are unlikely to report extreme working hours and low pay when filling out H-2A petitions for certification.
[78] *See* Department of Labor, *2015 Final Rule: Range Herding or Production of Livestock in the* United *States*, https://perma.cc/6PFE-CP65.
[79] Efraim Benmelech et al., Strong Employers and Weak Employers: How Does Employer Concentration Affect Wages?, National Bureau of Economic Research, WP 24307 (2018); Kevin Rinz, Labor Market Concentration, Earnings Inequality, and Earning Mobility, CARRA Working Paper Series, CARRA-WP-2018-10. (2018); Brad Hershbein et al., Concentration in U.S. local labor markets: evidence from vacancy and employment. data, Working Paper (2018); Yue Qiu & Aaron J. Sojourner, Labor-Market Concentration and Labor Compensation SSRN WP 3312197 (2019).
[80] See generally J.D. Matsudaira, Monopsony in the Low-Wage Labor Market? Evidence from Minimum Nurse Staffing Regulations, 1 Rev. Econ. Stat. (2013); D. Staiger, J. Spetz & L. Phibbs, Is There Monopsony in the Labor Market? Evidence from a Natural Experiment. 2 J. Labor Econ. 28 (2010).
[81] See Cahuc, supra note 7 at 793.
[82] See Julie L. Hotchkiss & Myriam Quispe-Agnoli, Employer Monopsony Power in the Labor Market for Undocumented Workers, Federal Reserve Bank of Atlanta, Working Paper 2009-14D (2009).
[83] Boris Hirsch & Elke J. Jahn, Is there Monopsonistic Discrimination Against Immigrants? ILR Review, 68(3), 501–528 (2012).
[84] Suresh Naidu et al., Monopsony Power in Migrant Labor Markets: Evidence from the United Arab Emirates, 6 J. Pol. Econ. 1735, 1742 (2016).
[85] See Eric M. Gibbons, Allie Greenman, Peter Norlander and Todd Sørensen, Monopsony Power and Guestworker Programs, 64 The Antitrust Bulletin 540 (2019).
[86] About 13% of the H-2A petition referenced in the data do not specify how they were filed. These statistics do not give us the full picture of what is occurring. It might be that undisclosed petitions are more likely to be from individual employers or associations or contractors but there is no way to tell.
[87] H-2A contractors are part of agencies that help individual employers recruit H-2A workers.
[88] José A. Azar, Ioana Marinescu, Marshall I. Steinbaum, and Bledi Taska, *Concentration in US Labor Markets: Evidence from Online Vacancy Data*, Labour Econ. (2020).
[89] *Id.*
[90] William M. Boal and Michael R. Ransom, *Monopsony in the Labor Market*, 35 J. Econ. Lit. 86 (1997).
[91] *Id.*
[92] *See* Azar et al., *supra* note 88.
[93] US Bureau of Labor Statistics, Standard Occupational Classification, https://perma.cc/4FJJ-PYC8 ("The [ ] Standard Occupational Classification (SOC) system is a federal statistical standard used by federal agencies to classify workers into occupational categories for the purpose of collecting, calculating, or disseminating data. All workers are classified into one of 867 detailed occupations according to their occupational definition. To facilitate classification, detailed occupations are combined to form 459 broad occupations, 98 minor groups, and 23 major groups. Detailed occupations in the SOC with similar job duties, and in some cases skills, education, and/or training, are grouped together.").
[94] *See* Eric M. Gibbons, Allie Greenman, Peter Norlander and Todd Sørensen, *Monopsony Power and Guestworker Programs*, 64 The Antitrust Bulletin 540 (2019).
[95] US Dep't of Justice & Fed. Trade Comm'n, Horizontal Merger Guidelines (2010).
[96] José Azar, Ioana Marinescu, and Marshall Steinbaum, *Labor Market Concentration*, 55 J. Human Resources 4 (2020).
[97] José Azar, Ioana Marinescu, and Marshall Steinbaum, *Labor Market Concentration*, 55 J. Human Resources 10 (2020).
[98] *Id.*
[99] *See* Gibbons, *supra* note 85.
[100] *Todd v. Exxon Corp.*, 275 F.3d 191, 208 (2d Cir. 2001).
[101] José Azar, Ioana Marinescu, Marshall Steinbaum, *Labor Market Concentration*, J. Human Resources (2020) (published online).
[102] *See* Bruno, *supra* note 12 at 22 (explaining that the changes in wage requirements were "highly controversial.").
[103] Jeffrey S. Passel and D'Vera Cohn, *U.S. Unauthorized Immigrant Total Dips to Lowest Level in a Decade*, Pew Rsch. Ctr. (Nov. 27, 2018), https://perma.cc/Y5TA-FHSH.
[104] *See supra* Part III.
[105] *See, e.g.*, Lindsey Lowell, *Growing Modern American Guestworkers: the Increasing Supply of Temporary H-2A Agricultural Workers*, IZA Discussion Paper *12 (2011).
[106] Selected statistics of top states granting H-2A labor certifications (in parentheses) in 2019: Florida (33,598), Georgia (29,480), Washington (26,226), California (23,321), North Carolina (21,605).

[107] *See* Marianne Levine, *Behind the Minimum Wage Fight, a Sweeping Failure to Enforce the Law*, Politico (Feb. 8, 2018), https://perma.cc/V8E5-4TUP.
[108] I initially thought I could work with Quarterly Workforce Indicators (QWI) which are a set of economic indicators including employment, job creation, earnings, and other measures of employment flows. The QWI are reported based on detailed firm characteristics (geography, industry, age, size) and worker demographics information (sex, age, education, race, ethnicity) and are available tabulated to national, state, metropolitan/micropolitan areas, and county. I computed the data but analysis was deeply flawed: on the one hand I had individual and employer level data for H-2A workers and on the other hand I had aggregated data for labor market indicators in an entire county which might or might not have included H-2A workers. The readily available data was too aggregated (by county or city) which did not allow for a granular analysis (especially necessary for the difference-in-differences estimates), and did not disentangle whether workers were US workers or other workers with a visa. Estimates were collinear.
The better data is the Quarterly Census of Employment and Wages (QCEW) and Longitudinal Database of Establishments (LDE). I am currently applying for this data (for this paper and another paper project), but it takes months.
[109] The data before 2014 is formatted differently than the 2014, 2015, 2016, 2017, 2018, and 2019 data. It will require a bit more cleaning and formatting before it can be merged with my current dataset. This is why I haven't been able to test the parallel trend assumption yet.
[110] *See e.g.*, THE WORLD BANK, *Difference-in-Differences*, https://perma.cc/QKK9-UTA2; Jonnathan Mummolo, *Causal Inference: Difference-in-Differences*, Stanford, https://perma.cc/7DWH-TNL3.
[111] *See* 20 C.F.R. § 655 at 62960 (detailing the comments received by growers and herders during the notice and comment period).
[112] Edward Kosack, Guest Worker Programs and Human Capital Investment: The Bracero Program in Mexico, 1942–1964, 56 J. Human Resources 570, 574 (2020).
[113] Id. (citing Gibson & McKenzie, The Development Impact of a Best Practice Seasonal Worker Policy, 96 Rev. Econ. Stat. 229 (2014)).
[114] Id. (citing Reichert & Massey, Guestworker Programs: Evidence from Europe and the United States and Some Implications for U.S. Policy, 1 Pop. Rsch. Pol. Rev. 1 (1982)).
[115] Id.
[116] And I assume that the consumer/domestic worker's taste are the same (she does not re-rank her preferences).
[117] See Cahuc, supra note 7 at 13–23.
[118] This is definitely not the case with the current H-2A program since the DOL requires that employers show a labor shortage for the same occupation.

# XI. LERA Annual Reports

## LERA Executive Board Meeting Minutes
## Thursday, February 10, 2022, 12 noon Eastern Time
## Via Videoconference

*Call to Order*—The meeting was called to order at 12:01 p.m. by Wilma Liebman, President. Present at the meeting were board members Daniel Altchek, Peter Berg, Julie Farb, Janet Gillman, Shannon Gleeson, Kati Griffith, Quenton Herbert, Erin Johansson, Tamara Lee, Javier Ramirez, Christine Riordan, Palak Shah, and Marc Weinstein. Board members and RVP's present were Beverly Harrison and Jim Pruitt. Committee Chairs present were Virginia Doellgast, Joel Cutcher-Gershenfeld, Abigail Wozniak, Harry Katz, Taz Hurst, David Lewin, Bonnie Castrey, and Steve Sleigh. Officers present included Ryan Lamare, Bill Canak, Adrienne Eaton, Paul Clark, Wilma Liebman, and Andrew Weaver. Staff present included both Emily Smith and Bernadette Tiemann.

*Approval of the Minutes*—The motion to approve the minutes passed unanimously for both the board meeting in June of 2021 as well as the General Membership Board Meeting in June of 2021 with no revisions. Bill Canak moved the motion to approved and Beverly Harrison seconded the motion which passed unanimously.

### Reports

*Discussion of Meeting Registration for Speakers*—Wilma Liebman, President, raised the question of charging speakers for meeting registration, specifically for practitioner speaker who intend only to attend the meeting to speak and participate at their session only and not attend other session on the meeting program. She reported that last year several meeting speakers questioned their need to pay for registration when their intention was to minimally participate in other ways. Wilma asked the board to consider alternative payment schemas, either to be considered on a case-by-case basis or some other system. After some discussion on the topic, it was tabled to hear from both the financial, membership, and program committees. After those reports were heard, the board continued discussion of this topic. Some ideas included: 1) a reduced one-day rate, perhaps $50, for virtual speakers, and/or 2) to publicize hardship scholarships for speakers. One board member thought that perhaps $50 was too low for a one-day rate for virtual conference registration and that we should focus on relaying the value of participating at the conference. Another board member offered that the goal of coming the meeting is building community and connection and dropping in to participate in only one session does not meet that goal. Many voiced support of hardship scholarships for those who could not otherwise financially participate but urged fiscal caution for enacting policies that the organization would not be able to financially sustain in the long run, considering that speakers far outweigh attendees. After divergent discussion, the topic was tabled for further examination and may again be considered at future board meetings.

*Financial Report*—LERA Secretary-Treasurer Andrew Weaver reports that had a slight decline in revenue in comparison to 2020. There is some softness around membership numbers which will discussed in more detail with the membership report and there are some strategies that have been proposed and built into the budget next year to address that. We had a small decline in revenue; however, expenses declined even more. We were able to realize an 11% net income up from a 6% net income in 2020, so there was in improvement there. Going forward, we expect our net income to again decline to 6% in 2022. There is currently high inflation, and we do need to make sure that our critical staff, Emily Smith and Bernadette Tiemann, who are doing an excellent job, are keeping pace. Sadly, we will not be increasing Emily's salary enough to keep pace with inflation, but the 2022 budget calls for a 3% raise for Emily Smith and a 6% raise for Bernadette Tiemann. We are also seeing higher expenses in publishing and mailing costs in 2022. With flagging membership, we are also increasing our membership promotion budget including direct mail and digital promotion to gain back necessary members to sustain the organization. All of this is going to create a tighter budget in 2022. Staff has done a good job holding the line and preventing pandemic-related losses as we struggle with not having in-person meetings and taking a more remote approach to the entire organization. We haven't seen the bottom fall out of membership and we've kept the organization together. Revenues from chapters have been really encouraging; there's been good chapter engagement. However, membership is dropping, and that is going to require investment and digital recruitment going forward. All of this is going to squeeze; that plus inflation over the next year will put our bottom line back into single digits in the budget for 2022. For the virtual meeting, we realized $104,000 in meeting revenue, which was right where we predicted we would. Contributions held steady from individuals at $35,000. The diversity and inclusion fund is comprised of two components: $10 from each full meeting registration on a four-year moving average and contributions to the Gershenfeld fund. Those two figures combined is about $9,067. Last year, after it was announced that the Gershenfeld fund was being repurposed to support diversity and inclusion our members responded generously with over $5,000 in contributions, up from $810 in the previous year. Emily Smith, LERA Executive Director, encourages the board to use the funds in ways that both meet the mission of increasing diversity and inclusion at LERA meetings and support the operating expenses of the organization, since the funds have been diverted from operating expense capital, for sustained financial stability. Tamara Lee motioned to accept the financial report and Bill Canak seconded the motion. The motion to accept the financial report from 2021 and the 2022 budget passed unanimously.

*Membership Committee Report*—David Lewin reports that Tazwell Hurst was recently appointed the second co-chair of this committee. The pandemic year one, 2020, was ironically good for membership, as membership increased. But 2021 membership is lower than 2020, but higher than 2019. In early 2022, a lag has begun, but we will work to increase membership throughout the year with a multi-faceted marketing promotion approach. The proportion of members that are represented by attorneys, government, management, and unions., has declined by about 10% since 2010. Neutral members have remained about the same. Academics, however, have increased their numbers. David Lewin makes this observation as a concern about its impact on practice point of view. He does not believe that LERA will ever be exclusively an academic organization but feels it's important to point this trend out. Wilma Liebman asked the question of what we should be doing to attract more members in the practitioner community. Adrienne Eaton had initiated some discussion last year, and there is a template letter included in the board packet that can be used by anyone on the board or membership committee as template language to reach out to members of management to invite them to join, and all members are invited to do so. Harry Katz mentions that it is important to continue connecting national members to our local chapters. Paul Clark mentions that Steve Greenhouse has been writing about the renewal of the "labor beat" among journalists, and that perhaps we could speak to Greenhouse about getting labor journalists more involved in LERA, especially in light of the fact that journalists can apply for press passes to our annual conferences. It was mentioned that Greenhouse will be involved at the next meeting, and there is another initiative in the works which may create a Labor Journalism Interest Section within LERA. At the LERA Presidential Address planned in 2022, Josh Eidelson, a Bloomberg reporter, will interview Wilma Liebman. Erin Johansson reports that unions may be interested in becoming more connected through the industry councils with LERA and that she will connect with Joel Cutcher-Gershenfeld on how to better implement this connection. Bill Canak thanks Harry Katz for bringing this up the important connection between national LERA and our chapters. Taz Hurst suggested we reach out to organizations like the AFL-CIO and ask them to promote LERA membership with LERA to their members. He mentioned the same strategy could be taken with other organizations he is familiar with including the CBTU APRI and the Machinists Union. Taz Hurst attends these conferences annually and may be able to promote LERA at those events. Tami Lee volunteers to ask members to find out their practitioner experiences of the meetings and of membership. Julie Farb comments that it might be effective to reach out to a group of international union research directors and invite them to connect with LERA. The AFL-CIO convention is coming up in June 12, 2022 in Philadelphia, and Julie Farb suggested this conference may be an excellent opportunity for LERA to have some sort of presence and promotion at the meeting. Wilma would like to follow-up with Julie Farb on these ideas. Taz Hurst mentions that the Machinists Union is having their conference in September. They have a training center in Southern Maryland, and they could reach out to their members though this venue. Emily Smith will follow-up with Taz following the Board meeting on how to connect Tax with LERA resources when he attends these conferences. Bonnie Castrey will work to connect local chapters with this effort. Erin Johansson asked about bringing back the National Policy Forum in DC. Wilma promises to work on these efforts with a goal to see increased membership by our June meeting 2022 particularly with practitioners. The membership committee report was adopted unanimously by the board. Peter Berg moved and Bill Canak seconded the motion.

*LERA Annual Meeting Program Committee Report*—Paul Clark reports that he is excited about the upcoming program and thanks the Vice Chairs Bill Spriggs, Bob Chiaravalli, and John Budd and the program committee members for their work. The theme of the program is "Elevating Voice and New Voices in the Workplace and Beyond." There were several sessions submitted on the meeting theme, and there are several strong plenaries planned, and one or two more that may still be announced. We are hoping that we get great attendance. Wilma Liebman thanks Paul Clark and his committee for such good work.

*ILERA/LERA Program Committee Report*—Harry Katz reports that he and Emily Smith are close to securing a contract with midtown Manhattan Hilton for the joint 2024 LERA/ILERA event that will take place the last week of June 2024. He is committed to creating a robust scholarship fund to support individuals to participate who could not otherwise from around the world including Southeast Asia, Africa, and Latin America. The initially proposed location in California did not pan out with the hotel, and the hotel in Manhattan not only has excellent facilities for our meeting but also seems to be offering us fairly preferential terms. We will report back at the next board meeting. We hope to attract 600-700 people at this exciting joint event.

*LERA@ASSA Program Committee Report*—Abigail Wozniak reports LERA@ASSA conducts 18 sessions in conjunction with the AEA/ASSA Annual Meetings, a tradition for LERA going back 74 years. In fact, LERA is one of the six founding members of the ASSA. These 18 sessions represent additional opportunities for LERA members to present their research and keep the issues of importance to LERA front and center in mainstream economics and with policy-makers. The 2021 program featured several important sessions with key speakers. To ensure that we have a sustainable model at this meeting, LERA requires that all those who wish to present a paper in a LERA session become a LERA member. Brad Hershbein was recently appointed as the second co-chair by Wilma Liebman. This year the committee has been re-furbished and is comprised of 16 strong and diverse committee members; it's an excellent committee. March 15th, 2022, is our session proposal deadline for our 2023 meeting in January; you are invited to submit your session proposals. Upcoming meetings are still all planned to be place-based meetings, though that could change. These include: 1) January 6-8, 2023 (Friday, Saturday & Sunday) New Orleans, LA, 2) January 5-7, 2024 (Friday, Saturday & Sunday) San Antonio, TX, and 3) January 3-5, 2025 (Friday, Saturday & Sunday) San Francisco, CA.

*Development Committee Report*—Harry Katz reports that development funds have contributed to the solid financial state of LERA reviewed by our secretary-treasurer. We raised $35,000 last year in individual contributions, and $40,000 from our major sustaining sponsors, Kaiser Permanente and Ford Motor Company. International Union, UAW is a new sustaining sponsor for LERA beginning in 2022 for $25,000. We are also building a quasi-endowment fund focusing on major gifts and final gifting, so please think about this in your estate planning. We are seeking additional sustaining sponsors, and if you have leads, Harry Katz and Jim Pruitt are happy to make the ask. Jim Pruitt made a personal request that each individual board member and officer consider donating a $1,000 each year, depending upon individual personal ability to do so. Jim also believes there could be interest from Kaiser Permanente managers in LERA membership and will connect with Emily Smith following the meeting to discuss further.

*Nominating Committee Report*—Steve Sleigh reported his slate of 31 potential candidates to the board for the June 2022 LERA election. His full committee, save Kris Rondeau who was unable to attend, met on Friday, February 4, 2022, and determined the slate that is contained in the board packet. The committee well-represented all the viewpoints of the organization. Wilma Liebman asked the board if they had any questions or had any comments after reviewing the slate and relayed that the board does have the authority to make changes to the slate before approving it. Liebman raised questions she had related to the slate of candidates: 1) two nominees in the neutral category are perhaps too similar to each other, and 2) a nominee in the management category is potentially being too busy accept. The slate passed unanimously without revision. Marc Weinstein motioned to approve the slate, and Bill Canak seconded the motion and it passed unanimously.

*Interest Sections/Industry Councils Coordinating Committee Report*—Joel Cutcher-Gershenfeld, reported the numbers of members who identify with the various LERA Interest Sections and Industry Councils. In this way, members are voting as to where their interests lie. Our challenge is how to appreciate that and how to expand it going forward. Momentarily, Paul Clark will present a new interest section charter for your review on pedagogy and teaching in our field, which we are all excited about. The second initiative is that there's organizing going on for a labor journalism/media industry council. It's still under discussion and yet to determine a name, but there are currently eight or nine alt media outlets that are unionized including some very well-known ones. And there are labor journalists still at major media outlets. It's likely that at the next annual meeting of the board, we may have a charter of that group coming. If you are interested in either of these initiatives, please contact Joel Cutcher-Gershenfeld, and he can point you to the individuals working on them. The third initiative is that we have identified the lowest-populated industry councils and see if we might be able to combine some of them and perhaps breathe new life into them. The real aim is for LERA to have communities of practice by interest and industry/sector, by which people can of course contribute to the program, contribute to webinars, and frankly be a voice for those industries or sectors in society. In the areas of construction, healthcare, and higher education, each have affiliates with other centers that deal with employment issues in those respective arenas, and that has been a powerful additional organizing feature.

*Teaching Interest Section Report*—Paul Clark reported that the idea of a LERA Teaching Interest Section arose at the June 2021 Annual Meeting. The Section organized a session on the program which reviewed innovation in the field and online teaching, and there was a good turnout. The group is continuing to move forward and it has two sessions on the upcoming June 2022 program. One of the things they did in preparation for the upcoming meeting was to send out a survey, asking people about the innovative things they have been doing in the classroom. Antone Aboud has been leading that, and Karla Katz has also been involved; they are both working through the responses to the survey. There has been plenty of interest, and those interested have shown initiative and follow-through, so I propose we take this forward. A motion to approve the Teaching Interest Section charter came from Joel Cutcher-Gershenfeld, was seconded by Adrienne Eaton, and was passed unanimously by the LERA Executive Board.

*Diversity and Inclusion Report*—Virginia Doellgast reported that we need to be making sure we are more effectively using the funding available to fund the people who need it to participate at LERA Annual Meetings. In trying to find a happy medium between making sure people have access, and not over-promising what our meager funding can afford, we have landed on placing information about diversity and inclusion funding for under-represented minorities on the session proposal submission forms for organizers to see when they submit their session proposals. These organizers are in the best seat to accurately assess and point us to speakers who may have a financial need and can alert the program committee and the diversity and inclusion committee to this. This information was not in the 2021 session proposal forms, but it is in the 2022 session proposal forms and will be there going forward. We still need to get this information reported to the diversity and inclusion committee for 2022 so they can review and approve any requests that have been made by session organizers. Additionally, there is another, stand-alone form that individuals can independently access, outside of the session proposal form, to request registration support. In 2021, 4 requests were made, and 4 requests were approved by this committee through that independent form. In 2022, no requests have yet been received. The committee has implemented a code of conduct that all meeting attendees are required to agree to, and meeting attendees are asked to report on their gender and race. Members are also asked to report on their gender and race. LERA has instituted language that specifies that those accepting awards must be free of harassment findings within a certain time period. Program Chairs are asked to consider diversity in their programs. In tracking sessions on the program that have a significant focus on equity and justice, based on race, gender, disability, and age, Doellgast reported 11 significant panels in both 2021 and 2022 which is an increase from prior years. President Liebman thanked the committee for their report and requested that they keep up the good work.

*Editorial Committee Report*—Ryan Lamare reports on the 2022 Research Volume "Industrial Relations and a Racial Reckoning" by Tamara Lee, Maite Tapia, Naomi Williams, and Sheri Davis Faulkner. This book is finishing up production with the majority of chapter in the production process. We anticipate being able to complete the volume and have everything ready to go and in the mail on time, and he thanked the editors for all their outstanding work. The 2023 Research Volume proposal is contained in the board packet for review "The Evolution of Workplace Dispute Resolution: An International Perspective," edited by Ariel Avgar, Cornell University; Deborah Hann, Cardiff University; Ryan Lamare, University of Illinois at Urbana-Champaign; and David Nash, Cardiff University. This proposed volume carries forward some trends that we would have seen through the LERA Research Volume process as far back as the late 90's where we had a volume on dispute resolution and workers' rights, edited by some folks here. We wanted to try to propose a volume that carries forward that conversation about the role of conflict in societies with a broader set of perspectives and twenty years later, considering cross-national and international components and also conversation about the role that conflict plays with major events and shocks that have occurred in the last twenty years, in particular the pandemic, the rise of #MeToo, and the role that conflict resolution programs play within the field of labor and employment relations. This proposal includes a variety of perspectives, international in particular. The editors have solicited chapter contributions that will go beyond the domestic focus that we tend to have in conflict research, and we have been able to secure chapters from authors focusing on the US, UK, Germany, Ireland, Australia, New Zealand, and others. Tamara Lee motioned to accept the proposal, and Beverly Harrison seconded the motion, and the motion passed unanimously.

*National Chapter Advisory Council Report*—William Canak, NCAC Chair, reports that after two years of pandemic-related challenges the LERA chapters are in exceptionally good shape. Many are up for chapter merit awards this year, and as you heard earlier, chapter dues payments have been great, which he believes is a mark of the health of local and state chapters. Thanks to Paul Clark, we have begun an initiative aimed at founding new chapters in several promising cities and states around the U.S. That initiative will be coming in the months ahead and we hope to give you progress updates in June this year. Many of you have seen evidence of chapter activity reported in the LERA eBulletins. This reporting confirms the energy and vitality of the local chapters. During the pandemic, the ways that they have innovated means they are engaging and serving their members. We have RVP election coming up for the East region. I'll note that all regional chapter vice presidents have served with distinction and really confirmed the excellence of this change to LERA's bylaws a number of years ago. In addition, LERA chapter liaisons have been reassigned this past year and are working, and are available to work, with chapter RVP's and individual chapters. The NCAC has had several virtual meetings in the last year which were engaging and productive. In June, we will conduct meetings of the NCAC, a chapter representatives meeting, and we'll again conduct our chapter administration workshop session. This year's session will feature Virginia LERA, a newer but highly productive chapter, and COLERA, which has been extremely active during the pandemic. Each chapter will talk about their approaches. Bill Canak ended his report by encouraging all LERA national members, and board members, to join their local chapter, as required by LERA bylaws. If anyone needs help connecting with a local chapter, Bill stands by to assist.

Next LERA Executive Board meeting will take place virtually on Wednesday, June 1, 2022, at 6-7:30 p.m. Eastern Time (New York Time).

*Adjournment*—The meeting was adjourned at 12:52 p.m. Eastern Time by President Wilma Liebman.

# LERA Executive Board Meeting Minutes Wednesday, June 1, 2022, 5:30–7 p.m. Eastern Time Via Videoconference

*Call to Order*—The meeting was called to order at 5:30 p.m. by Wilma Liebman, President. Present at the meeting were board members Daniel Altchek, Peter Berg, Julie Farb, Janet Gillman, Shannon Gleeson, Kati Griffith, Quenton Herbert, Erin Johansson, Tamara Lee, Javier Ramirez, Christine Riordan, Palak Shah, and Marc Weinstein. Board members and RVP's present were Beverly Harrison and Jim Pruitt. Committee Chairs present were Virginia Doellgast, Joel Cutcher-Gershenfeld, Abigail Wozniak, Harry Katz, Taz Hurst, David Lewin, Bonnie Castrey, and Steve Sleigh. Officers present included Ryan Lamare, Bill Canak, Adrienne Eaton, Paul Clark, Wilma Liebman, and Andrew Weaver. Staff present included both Emily Smith and Bernadette Tiemann.

*Approval of the Minutes*—The motion to approve the minutes passed unanimously for board meeting in February of 2022 with no revisions.

*LERA 76th Annual Meeting Update*—Harry Katz reports that the meeting that will occur in 2024. Harry began visiting this board a few years ago to coordinate the ILERA triennial world congress synchronously with the LERA 76th Annual Meeting. The latest development is that Javier Ramirez from the FMCS proposes that in 2024 we also coordinate this meeting jointly with the FMCS National Labor-Management Conference. This meeting is held every other year and draws a substantial crowd of labor and management practitioners. The board discussed, offered their views, and expressed enthusiasm for this joint venture. The venue for the LERA 76th Annual Meeting, the Hilton Midtown Manhattan, is of a scale that will accommodate all three organizations. The plans are underway, the date is June 26-30, 2024, and the event should be engaging with something to appeal to all constituencies. There will be particular events that will be still occur, like General Membership Meeting for LERA and ILERA, Presidential Luncheon for LERA, there is plenty of session space and common space for attendees to interact and network. Javier Ramirez at the FMCS reports that LERA needs to both maintain academic elements of the conference and to generate revenue. Neither of these issues are concerns for the FMCS. The FMCS does have their conference every two years and would like to have this collaboration recur every other year, with a memorandum of understanding between our two organizations. This gives us a significant opportunity to increase LERA membership, drive revenue for LERA, etc. There are many opportunities for efficiency and redundancy reduction, including: key speakers (such as the Secretary of Labor Marty Walsh), contributors who would need to contribute to a single event, shared administration and meeting platforms, etc. We have the ability to work closely on the content that overlaps and keep other content separate. The main question is: Why have we not done this sooner? Here is a chance to do it, and even if we experience a learning curve, and we will continue to improve the process for 2026 and beyond, and this this the main event to attend in the field of LER. At this point we are committing to having a joint meeting in 2024. If all goes well, we can continue doing this in 2026 and going forward. Harry Katz mentions that LERA has many academics, and FMCS has many practitioners, and this gives us an opportunity to enhance the experience of the parties involved on both sides, somewhat broader to where LERA operated 20-30 years ago. If we have more practitioners at the meeting, it can only elevate the event. To tighten the bond even further, Joel Cutcher-Gershenfeld suggests that we might have FMCS mediators tied to various LERA Industry Councils and have sessions at the meeting featuring the mediators. The board voted and the vote was unanimous to move ahead with the joint collaboration. Additionally, a memorandum of understanding will be developed between our two organizations (LERA and the FMCS) to formalize this relationship.

## Reports

*Financial Report*—LERA Secretary-Treasurer Andrew Weaver reports that the final 2021 audit has not been completed but we should be getting the final 2021 numbers soon. This is hopefully the last year of our virtual conference. Revenues declined a small amount, but expenses declined by more. The bottom line is that we were in the black in 2021 and we cleared a percentage of net income that is consistent of our past experience under Emily Smith's directorship. On the whole, 2021 ended up in a good place. Going forward, the revised 2022 budget, we have had a very good story on sponsorship thank you to UAW, Ford, and Kaiser Permanente and the Alliance of Healthcare Unions. We do need to make some investments. First, salaries increased by one percent from 2021 to 2022. In coming years, we may look at increasing salaries. There is extra investment in the 2022 budget for extra-help in the fall during the busy season. We also have investments planned for our library digital subscriptions, and chapter websites, and other matters. In 2022, we are expecting to clear about 6% net income or a little over $20,000 and stay in the black. On the whole, LERA is in good shape, and when we ratchet back up to in-person conferences we will be in a good footing. The LERA board accepted the financial report.

*Membership Committee Report*—David Lewin reports that membership numbers were flagging in 2021, but that those numbers have been shored up in 2022 due to the new major sustaining sponsors, which carries a component of membership as well. You will notice that the price of membership has stayed the same for about a decade, which is remarkable, especially when benchmarked to other organizations. We have held the line on dues. This, and our positive financial report, is all do to a constrained financial approach. Wilma Liebman says that her regret over the last year has been being unsuccessful with recruiting more practitioners in LERA, and she would love to see this portion of the membership grow. Perhaps the collaborative event planned for 2024 will help this. Kevin Legel has begun publicizing an opportunity for their staff to have discounted memberships at LERA through Ford's major sustaining sponsorship.

*LERA Annual Meeting Program Committee Report*—Bill Spriggs reports that due to technical glitches, today's earlier program committee meeting will be repeated in late June or early July. The theme for the LERA 75$^{th}$ Annual Meeting, June 1-4, 2023, in Detroit, Michigan is planned to be: "Workers and American Democracy: Challenges Ahead, Lessons from Before." Bill shares that he has been personally troubled by the demise of democracy globally, and he encourages papers or sessions where we can appreciate this, and think about contributions of labor to the legal democracy in the United States, its impact on American workers and global workers and on American foreign policy, etc. We hope to examine the attacks on democracy and related challenges, but also to look to the future and point to people who are engaging in collective action and education around organizing. We are currently in a moment where new types of collective action are taking place, and it is appropriate to pay attention to historical contributions of the labor movement to the quality of work, life, and employee involvement, and hypotheses that if people had more of a say at work, it would roll over into civic life and improve democracy, an idea which has been supported by a number of studies.

*Development Committee Report*—Harry Katz reports that LERA is in good shape on the development front as we now have three distinct major sustaining sponsors including UAW, Ford Motor Co., and Kaiser Permanente and the KP Alliance of Healthcare Unions. Please know that when you and others make contributions to LERA, you can always target gifts to specific worthy causes, such as various funds. We are also building a quasi-endowment fund focusing on major gifts and final gifting, so please think about this in your estate planning or major gift. We are seeking additional sustaining sponsors, and if you have leads, Harry Katz and Jim Pruitt are happy to make the ask. Jim Pruitt made a personal request that each individual board member and officer consider donating a $1,000 each year, depending upon individual personal ability to do so.

*Interest Sections/Industry Councils Coordinating Committee Report*—Joel Cutcher-Gershenfeld reported that the industry councils and interest sections will meet Friday at 3 pm and they will take a deeper dive into the information that is contained in the report attached in the board packet. Joel will be asking the section and council leaders about potential rebranding. Secondly, we have had low involvement in the airline and utility councils and those may need to be retired. Joel looked through the program with an eye towards content about what sessions speak to issues of interest to various sections and councils, and there were several sessions of interest to many of them. At our meeting on Friday, we will be discussing the possibility of exploring new sections or councils on the areas that were most noted on the program. The current leadership on councils and sections are listed in the report and we would love to correspond a federal mediator to each of these groups. The media award committee has been working to establish a labor journalism group and there will be a panel discussion by former award winners on the program this year. We feel we are at a point at reaching out to these individuals and establishing a formal network of labor journalists. Quite a few of these labor journalists are organizing themselves into unions. Steven Greenhouse has a list of a few dozen labor journalists that he is in touch with and he is willing to share this list with the group. Javier Ramirez looks forward to having that conversation about mediators and also has some suggestions about the airline industry council. When people speak to labor journalists, they should be given the heads up about LERA and tell them about the new journalism group. The newest interest section is on teaching and pedagogy, which Paul Clark has played a large role in creating.

*Diversity and Inclusion Report*—Christine Riordan reports that we have added in gender and ethnicity demographics into multiple touchpoints of LERA processes, and we have collected about half of our members' demographics and about two-thirds of our meeting registrant's demographics. One way in which we are currently using this is to work with the program committee. Organizers have the ability to give information to the program committee on the diversity and inclusion of their session. We also have a fund to build diversity and inclusion in various ways. We have begun to lay a foundation for collecting information and building these efforts into our processes. We are looking at giving membership to those who are approved for meeting registration. We are also looking at providing childcare resource information to LERA meeting attendees. The committee discussed LERA's meeting locations where issues of women's healthcare is de-prioritized. The committee is supportive of looking into this further and seeing what other organizations are doing and seeing whether any action by LERA is feasible. Harry Katz offered that the committee is not yet coming forward with a proposal to alter how LERA does our site selections, but please do keep in mind that LERA does strive to do two things: we want to be in a union hotel and we want to be in a location that has a local LERA chapter, and that he would caution against reducing our support or services to those chapters. Bill Canak expresses that he supports meeting in locations where LERA local chapters exist or are forming and that this should be an important consideration when reviewing where to conduct LERA meetings.

*Editorial Committee Report*—Ryan Lamare reports that the LERA/ILR Review process had ten submissions after the last June conference. Of those, three moved into the review process, and two remain in the review process. Later this summer, we will discuss how to make this process more efficient and get the papers through the process in a more compressed timeline. The LERA 2022 Research Volume "Industrial Relations and a Racial Reckoning" by Tamara Lee, Maite Tapia, Naomi Williams, and Sheri Davis Faulkner is almost out the door; we are proud of the volume and will be delighted to see it in print. Tamara Lee and Matie Tapia who are on today's call have done a tremendous amount on this book. This book will be on the shelves in the next few months. The 2023 Research Volume "The Evolution of Workplace Dispute Resolution: An International Perspective," edited by Ariel Avgar, Cornell University; Deborah Hann, Cardiff University; Ryan Lamare, University of Illinois Urbana-Champaign; and David Nash, Cardiff University was approved in February. For both of these volumes, there are sessions on the program to take a look at this content. The conflict resolution volume is moving along well. At our next meeting, we will likely ask for your approval of the 2024 research volume which will be an update of the 2013 research volume on collective bargaining under duress. This will be edited by Howard Stanger, Paul Clark, and John Delaney.

*National Chapter Advisory Council Report*—William Canak, NCAC Chair, reports that LERA chapters are in good shape. Chapters are active and coming out of the pandemic era, chapter dues payments have been great. We have new chapters that have formed (Virginia LERA), and thanks to Paul Clark, we have begun an initiative aimed at founding new chapters in several promising cities and states around the U.S. such as southern Florida and in Mississippi and Louisiana. That initiative will be coming in the months ahead and we hope to give you progress updates in June this year. Many of you have seen evidence of chapter activity reported in the LERA eBulletins. This reporting confirms the energy and vitality of the local chapters. We have RVP election coming up for the East region. I'll note that all regional chapter vice presidents have served with distinction and really confirmed the excellence of this change to LERA's bylaws a number of years ago. In addition, LERA chapter liaisons have been reassigned this past year and are working, and are available to work, with chapter RVP's and individual chapters. The NCAC has had several virtual meetings in the last year which were engaging and productive. In June, we will conduct meetings of the NCAC, a chapter representatives meeting, and we'll again conduct our chapter administration workshop session. This year's session will feature Virginia LERA, a newer but highly productive chapter, and COLERA, which has been extremely active during the pandemic. Each chapter will talk about their approaches. Bill Canak ended his report by encouraging all LERA national members, and board members, to join their local chapter, as required by LERA bylaws. If anyone needs help connecting with a local chapter, Bill Canak, Bonnie Castrey, and the NCAC members stand by to assist.

Some of the attachments include the LERA@ASSA 2023 program, the 2022 LERA award winners, the LERA 2021 election winners. Paul Clark reports that the program committee feels great about it's theme of Elevating Voice and New Voices in the Workplace for the current annual meeting. It was mentioned today that one of the beautiful things about the virtual meeting is that people can continue to register right up to and through the meeting. The opening session tomorrow is with the U.S. Secretary of Labor Marty Walsh. Wilma Liebman did invite him to come back next year in person. Wilma Liebman thanks the staff for their dedication. Deborah Mueller came to a conference a few years ago and left with a mission to regenerate the Western NY LERA (Buffalo) chapter, and then the pandemic changed everything. As things have gotten back to normal, she had other circumstances that have precluded this activity again. However, she can see the finish line and is hoping that we can get back to this goal of revitalizing this chapter and will send out some thoughts and suggestions to potential volunteers and reach out to Virginia LERA and others for suggestions and best practices on chapter-building. Further, Deborah reports seeing large labor strikes resulting in big changes for both workers and communities, and that this is a good time for the Buffalo Chapter to revive. Bill Canak encourages her to come to the chapter workshop featuring this exact topic and to additionally contact the NCAC for additional support.

Next LERA Executive Board meeting will take place virtually on February 21, 2023, at Noon Eastern Time, via video conference.

*Adjournment*—The meeting was adjourned at 5:51 p.m. Eastern Time by President Wilma Liebman.

# LERA General Membership Meeting Minutes Friday, June 3, 2022, 3:45 p.m. Eastern Time Via Videoconference

*Call to order*—Wilma Liebman opened the meeting at 3:45 pm Eastern Time, and introduced Paul Clark, LERA President Elect and at the end of the LERA 74th Annual Meeting, officially the LERA President.

*Financial Report*—Andrew Weaver, Secretary-Treasurer, reported that the LERA is in good shape financially since expenses were held down in relation to decreased income in 2021 and 2022. However, we are looking forward to returning to the full financials from pre-pandemic era with the return of in-person meetings in 2023.

*Membership Committee Report*—David Lewin, Co-Chair, reported membership is in good shape. Academics comprise about 55% of our membership and other groups account for the rest. Given the registration for this particular conference, which is around 560 people, Lewin felt we could look forward to additional membership increases. This 2021 increase is due to a new major sustaining sponsor, United Auto Workers, and our other major sustaining sponsors (Kaiser Permanente, Ford, etc.) If we were to count local chapter members in our numbers, we would be closer to around 4,000 members, of course.

*Special Announcement*—Javier Ramirez and Harry Katz. Harry Katz is currently the president of ILERA in the midst of a 3-year term. At the end of his term, he will organize a world congress of ILERA which will happen in the US. Two years ago, it was decided that ILERA and LERA would conduct our meetings synchronously, and most recently Javier Ramirez suggested that we also synchronously conduct the FMCS national labor-management conference. The LERA Executive Board agreed to that as did the ILERA executive board, and so we are now planning a meeting the last week of June 2024 which will lead to an enriched program, and it should help expand LERA membership among the practitioner community.

Javier Ramirez had approached LERA and LERA responded that they needed to keep our sessions for our academics. With the FMCS, they run into issues with money every year, and it also benefits with LERA, and the thought is that every two years we could continue to meet together. We will make sure there are protections for LERA in place with a memorandum of understanding in case future administrations decide to move in a different direction. In the future, the FMCS will be collaborating with us, and there are overlays in capacities with similar content. And, it's a way to build the membership for LERA, there are pricing structures that would give discounts to LERA members, and it makes sense for the labor-management community at large. Networking is going to increase, and partnership between academics and practitioners will increase. The meeting will take place at the Midtown Manhattan Hilton on 6th Street, and we will be able to socially interact and network as community.

*Development Committee Report*—Harry Katz and Jim Pruitt, Co-Chairs, reported contributions are healthy as are our finances more generally. As was already mentioned by David, we benefit from having three distinct contributions (one from UAW, one from Ford, and one from KP and the KP Alliance of Health care Unions). We also have solid annual fund-raising drive. Keep in mind you can give money in either a restricted manner, or something that is dear to your heart like helping graduate students attend these meetings, or for worker centers attend our meetings. Jim Pruitt reports that this general membership meeting is very positive and makes a long-term LERA members feel that we are moving in a positive direction. Jim asked anyone in our general membership to contact the LERA staff and make a personal donation.

Rose Batt weighed in that the diversity within this organization has really grown and that she had the good fortune of not knowing most of the people at this conference, and additionally feels like there is a real momentum at the organization. There are all sorts of creative things that we can do to keep this thing rolling. Rose was impressed by what she has seen and appreciates the changes and differences. Wilma Liebman mentions that some of it is owed to the unique dynamism of the moment that we are in currently.

*Editorial Committee Report*—Ryan Lamare reports that the LERA/ILR Review special section had ten papers initially submitted and that there were two still in the running. For this year, please submit your paper by June 30, and Ryan promises to try to get an expedited review as possible. A couple of other updates, LERA's 2022 Research Volume "Industrial Relations and a Racial Reckoning" will be released shortly and the editorial committee is proud of this book. Our 2023 Research Volume will be on conflict resolution and will highlight the current system of conflict management in various countries and across various industries. That will be edited by Ariel Avgar, Debra Hahn, Ryan Lamare, and David Nash, and there will be a session on the program tomorrow highlighting these topics.

*Program Committee Report*—Bill Spriggs, LERA President Elect and Program Committee Chair reports that for the LERA 75th Annual Meeting, the theme will be workplace democracy, a topical and timely topic, especially with the concerns worldwide about the erosion of democratic institutions. The meeting will take place in-person in Detroit, Michigan in early June. Please submit your session proposals to the program committee by November 15, 2022.

*Diversity and Inclusion Committee Report*—Christine Riordan reports that the Diversity and Inclusion committee has been working to promote diversity and inclusion within our membership and at our meetings. Our sexual harassment policy is part of that, and last year we instituted a number of ways to collect demographic information on our meeting attendees and with our members. We have been working with the program committee to increase diversity of speakers. The Gershenfeld fund has been repurposed to support diversity at LERA annual meetings. There are a number of ideas that the committee is considering for future work.

*National Chapter Advisory Council Report*— Bill Canak, Chair, and Bonnie Castrey, Vice Chair of the National Chapter Advisory Council report that the state of the chapters is excellent. We have a number of chapters receiving chapter merit awards and outstanding chapter recognition. We are working with the FMCS to put together an organizing committee in Miami-South Florida area. We are also working with the FMCS to establish a new chapter in the southern Mississippi/New Orleans area. Those of you who read the bi-monthly LERA e-newsletter, please note that LERA is covering LERA chapter events, which helps us become a more cohesive organization. The vast majority of local chapter members are not academic, and the various activity that goes on in the local chapter, and you will see that activity is pragmatic and practitioner-oriented. Many of the changes made in the organization help support this increased weight of practitioners in the organization. We have two candidates running in the election this year for the East RVP. We have three chapter events happening at this conference this year, and much of the credit for these being successful and well-attended goes to the LERA staff and to the chapters. The chapter administration workshop will feature COLERA and Virginia LERA, our newest chapter. Since the meeting in 2016 in Minnesota, the bylaws of LERA have required all national members to be a member of a local chapter, and if you look into your own status and you see that you are not a local chapter member, please contact Bill Canak or LERA staff and they will help you find a local chapter. Wilma Liebman found it interesting and encouraging that LERA has chapters arising in the south. She further commented that LERA is a unique forum where all perspectives intersect.

*Industry Councils and Interest Sections Coordinating Committee Report*—Joel Cutcher-Gershenfeld, Chair, reported that we have eight established interest sections and one newly formed section, the Teaching Interest Section. We also have eight established industry councils and three new ones under discussion. Many of these groups are represented on the annual meeting program. When LERA members join or renew their membership they indicate which interest sections they want to affiliate with, and which industry councils, if any, with which they want to affiliate and he reported on current numbers in all these LERA groups.

*LERA Awards Committee Report*—Janice Bellace and Rose Batt, Co-Chairs, reported on the LERA 2022 Awards Winners. The capstone Lifetime Achievement Award was awarded to Bernard Anderson and Lisa Lynch. The Academic Fellows were Sarosh Kuruvilla, Chris Tilly, Françoise Carré, and Douglas Kruse. Practitioner Fellows were Meeta Bass, Erin Johansson, Mickey Brock, and Sheila Mayberry. The Susan C. Eaton Outstanding Scholar-Practitioner Award was conferred on Steven Greenhouse. The Myron C. Taylor Management Award was not conferred in 2022. The John T. Dunlop Outstanding Scholar Awards were given to Rachel Aleks and Chiara Benassi. The LERA Outstanding Practitioner Award was awarded to Owen Herrnstadt. The James G. Scoville Best International/Comparative IR/HR Paper Award was received by Chunyun Li. The Kenneth May Media Award was bestowed on Eli Rosenberg, Margot Roosevelt, Katie Johnston, and Michael Sainato. This year's Thomas A. Kochan and Stephen R. Sleigh Best Dissertation Award went to Debbie Goldman. The Susan C. Eaton Research Grant award was given to Jasmine Platt, and Ron Schatz. The DCDC Best Student Paper award winner was Jianxuan Lei.

*Future LERA Meetings*—To formally announce the dates of future meetings, we will be in Detroit in 2023 June 1-4, 2023 at the MotorCity Hotel and Casino. The LERA@ASSA 2023 will be in New Orleans, January 6-8, 2023, and the 2024 annual meeting will take place in New York City in Midtown Manhattan at the Hilton.

*Presidential Handoff*—Wilma Liebman formally handed over the gavel to Paul Clark, who will be assuming the mantle of LERA President at the end of this meeting. Wilma congratulated Paul on putting together this year's program. Paul Clark reported that it has been a pleasure to work with Wilma over the last year, and that she would be a tough act to follow. Paul shared that LERA has been a very important part of his professional life, and that he would his best to make sure that LERA is in good shape at the end of his presidential year.

*Last Announcements*—Tomorrow morning at 10 a.m. Eastern, LERA will kick off with the opening plenary on supply chains moderated by Tom Kochan followed by a full day of breakouts and sessions, and the presidential event which will feature our two lifetime achievement award winners.

# Labor and Employment Relations Association

Champaign, Illinois

Financial Statements

For the Years Ended December 31, 2022 and 2021

Feller & Kuester CPAs PLLC
Certified Public Accountants
806 Parkland Court
Champaign, IL 61821
217-351-3192

TABLE OF CONTENTS

**_Feller & Kuester CPAs PLLC_**

**Tax - Audit - Bookkeeping**
806 Parkland Court - Champaign, Illinois 61821
***Phone - (217) 351-3192 Fax - (217) 351-4135 Email - admin@fellerkuester.com***

INDEPENDENT AUDITOR'S REPORT

To the Board of Directors of
Labor and Employment Relations Association
Champaign, Illinois

## Opinion

We have audited the accompanying financial statements of the Labor and Employment Relations Association (a nonprofit organization), which comprise the statements of financial position as of December 31, 2022 and 2021, and the related statements of activities, functional expenses, and cash flows for the years then ended, and the related notes to the financial statements.

In our opinion, the financial statements referred to above present fairly, in all material respects, the financial position of the Labor and Employment Relations Association as of December 31, 2022 and 2021, and the changes in its net assets and its cash flows for the years then ended in accordance with the accounting principles generally accepted in the United States of America.

## Basis for Opinion

We conducted our audits in accordance with auditing standards generally accepted in the United States of America. Our responsibilities under those standards are further described in the Auditor's Responsibilities for the Audit of the Financial Statements section of our report. We are required to be independent of the Labor and Employment Relations Association and to meet our other ethical responsibilities in accordance with the relevant ethical requirements relating to our audits. We believe that the audit evidence we have obtained is sufficient and appropriate to provide a basis for our audit opinion.

## Responsibilities of Management for the Financial Statements

Management is responsible for the preparation and fair presentation of the financial statements in accordance with accounting principles generally accepted in the United States of America, and for the design, implementation, and maintenance of internal control relevant to the preparation and fair presentation of financial statements that are free from material misstatement, whether due to fraud or error.

In preparing the financial statements, management is required to evaluate whether there are conditions or events, considered in the aggregate, that raise substantial doubt about the Labor and Employment Relations Association's ability to continue as a going concern within one year after the date that the financial statements are available to be issued.

## Auditor’s Responsibilities for the Audit of the Financial Statements

Our objectives are to obtain reasonable assurance about whether the financial statements as a whole are free from material misstatement, whether due to fraud or error, and to issue an auditor’s report that includes our opinion. Reasonable assurance is a high level of assurance but is not absolute assurance and therefore is not a guarantee that an audit conducted in accordance with generally accepted auditing standards will always detect a material misstatement when it exists. The risk of not detecting a material misstatement resulting from fraud is higher than for one resulting from error, as fraud may involve collusion, forgery, intentional omissions, misrepresentations, or the override of internal control. Misstatements are considered material if there is a substantial likelihood that, individually or in the aggregate, they would influence the judgment made by a reasonable user based on the financial statements.

In performing an audit in accordance with generally accepted auditing standards, we:

- Exercise professional judgment and maintain professional skepticism throughout the audit.
- Identify and assess the risks of material misstatement of the financial statements, whether due to fraud or error, and design and perform audit procedures responsive to those risks. Such procedures include examining, on a test basis, evidence regarding the amounts and disclosures in the financial statements.
- Obtain an understanding of internal control relevant to the audit in order to design audit procedures that are appropriate in the circumstances, but not for the purpose of expressing an opinion on the effectiveness of the Labor and Employment Relations Association's internal control. Accordingly, no such opinion is expressed.
- Evaluate the appropriateness of accounting policies used and the reasonableness of significant accounting estimates made by management, as well as evaluate the overall presentation of the financial statements.
- Conclude whether, in our judgment, there are conditions or events, considered in the aggregate, that raise substantial doubt about the Labor and Employment Relations Association's ability to continue as a going concern for a reasonable period of time.

We are required to communicate with those charged with governance regarding, among other matters, the planned scope and timing of the audit, significant audit findings, and certain internal control related matters that we identified during the audit.

Feller & Kuester CPAs PLLC

Feller & Kuester CPAs PLLC
Champaign, Illinois

May 30, 2023

## LABOR AND EMPLOYMENT RELATIONS ASSOCIATION
## STATEMENTS OF FINANCIAL POSITION
## DECEMBER 31, 2022 AND 2021

| | 2022 | 2021 |
|---|---|---|
| **ASSETS** | | |
| **Current Assets** | | |
| Cash and Cash Equivalents | $ 435,876 | $ 394,971 |
| Investments | 81,978 | 85,955 |
| Accounts Receivable, Net | 1,773 | 1,406 |
| Prepaid Expenses | 18,219 | 5,626 |
| Inventory | 7,228 | 6,372 |
| Total Current Assets | 545,074 | 494,330 |
| Property and Equipment, Net | 1,630 | 3,292 |
| TOTAL ASSETS | $ 546,704 | $ 497,622 |
| **LIABILITIES AND NET ASSETS** | | |
| **Current Liabilities** | | |
| Accounts Payable | $ 250 | $ - |
| Accrued Liabilities | 45,077 | 39,661 |
| Dues Collected in Advance | 84,283 | 75,917 |
| Subscriptions Collected in Advance | 12,716 | 7,837 |
| Deferred Chapter Dues | 2,850 | 5,200 |
| Other Deferred Revenue | 10,450 | 54,550 |
| Total Current Liabilities | 155,626 | 183,165 |
| **Net Assets** | | |
| Without Donor Restrictions | | |
| Designated by the Board | | |
| Susan C. Eaton Scholar-Practitioner Memorial Fund | 16,779 | 20,779 |
| Kochan-Sleigh Best Dissertation Award Fund | 35,319 | 36,439 |
| Jim Scoville Award Fund | 2,176 | 2,796 |
| John T. Dunlop Public Policy Fund | 16,503 | 18,326 |
| Gladys and Water Gershenfeld Diversity & Inclusion Fund | 17,939 | 16,585 |
| Quasi-Endowment Fund | 45,373 | 45,462 |
| Total Designated by the Board | 134,089 | 140,387 |
| Undesignated | 222,085 | 145,194 |
| Total Without Donor Restrictions | 356,174 | 285,581 |
| With Donor Restrictions | 34,904 | 28,876 |
| Total Net Assets | 391,078 | 314,457 |
| TOTAL LIABILITIES AND NET ASSETS | $ 546,704 | $ 497,622 |

See Accompanying Notes.

## LABOR AND EMPLOYMENT RELATIONS ASSOCIATION
## STATEMENT OF ACTIVITIES
## FOR THE YEAR ENDED DECEMBER 31, 2022

| | Without Donor Restrictions | | | With Donor | |
|---|---|---|---|---|---|
| | Undesignated | Designated | Total | Restrictions | Total |
| **Support and Revenues** | | | | | |
| Membership Dues | $ 150,132 | $ - | $ 150,132 | $ - | $ 150,132 |
| Meeting Income | 115,145 | - | 115,145 | 5,888 | 121,033 |
| Organization Dues & Sponsorships | 94,227 | - | 94,227 | 550 | 94,777 |
| Subscriptions | 14,121 | - | 14,121 | - | 14,121 |
| Chapter Fees | 10,510 | - | 10,510 | - | 10,510 |
| Member Contributions | 48,299 | 3,775 | 52,074 | - | 52,074 |
| Royalties | 6,257 | - | 6,257 | - | 6,257 |
| Ad Income | 9,130 | - | 9,130 | - | 9,130 |
| Administrative Fees | 1,195 | - | 1,195 | - | 1,195 |
| Investment Income | (22) | (3,955) | (3,977) | - | (3,977) |
| Interest Income | 139 | - | 139 | - | 139 |
| Net Assets Released from Designation | 6,528 | (6,118) | 410 | (410) | - |
| Total Support and Revenues | 455,661 | (6,298) | 449,363 | 6,028 | 455,391 |
| **Expenses** | | | | | |
| Program Services | | | | | |
| General | 236,920 | - | 236,920 | - | 236,920 |
| Meetings | 2,709 | - | 2,709 | - | 2,709 |
| Publications | 40,516 | - | 40,516 | - | 40,516 |
| Supporting Services | | | | | |
| Management and General | 70,967 | - | 70,967 | - | 70,967 |
| Membership Development | 27,658 | - | 27,658 | - | 27,658 |
| Total Expenses | 378,770 | - | 378,770 | - | 378,770 |
| Change in Net Assets | 76,891 | (6,298) | 70,593 | 6,028 | 76,621 |
| Net Assets, Beginning of Year | 145,194 | 140,387 | 285,581 | 28,876 | 314,457 |
| Net Assets, End of Year | $ 222,085 | $ 134,089 | $ 356,174 | $ 34,904 | $ 391,078 |

See Accompanying Notes.

**LABOR AND EMPLOYMENT RELATIONS ASSOCIATION**
**STATEMENT OF ACTIVITIES**
**FOR THE YEAR ENDED DECEMBER 31, 2021**

| | Without Donor Restrictions | | | With Donor | |
|---|---|---|---|---|---|
| | **Undesignated** | **Designated** | **Total** | **Restrictions** | **Total** |
| **Support and Revenues** | | | | | |
| Membership Dues | $ 148,934 | $ - | $ 148,934 | $ - | $ 148,934 |
| Meeting Income | 104,141 | - | 104,141 | 5,645 | 109,786 |
| Organization Dues & Sponsorships | 56,029 | - | 56,029 | 550 | 56,579 |
| Subscriptions | 15,348 | - | 15,348 | - | 15,348 |
| Chapter Fees | 12,175 | - | 12,175 | - | 12,175 |
| Member Contributions | 35,174 | - | 35,174 | - | 35,174 |
| Royalties | 6,463 | - | 6,463 | - | 6,463 |
| Ad Income | 10,230 | - | 10,230 | - | 10,230 |
| Publications | 1,272 | - | 1,272 | - | 1,272 |
| Administrative Fees | 1,535 | - | 1,535 | - | 1,535 |
| Investment Income | (932) | (4,434) | (5,366) | - | (5,366) |
| Interest Income | 310 | - | 310 | - | 310 |
| Net Assets Released from Designation | 2,385 | (2,035) | 350 | (350) | - |
| Total Support and Revenues | 393,064 | (6,469) | 386,595 | 5,845 | 392,440 |
| **Expenses** | | | | | |
| Program Services | | | | | |
| General | 217,170 | - | 217,170 | - | 217,170 |
| Meetings | 10,120 | - | 10,120 | - | 10,120 |
| Publications | 36,522 | - | 36,522 | - | 36,522 |
| Supporting Services | | | | | |
| Management and General | 77,543 | - | 77,543 | - | 77,543 |
| Membership Development | 13,648 | - | 13,648 | - | 13,648 |
| Total Expenses | 355,003 | - | 355,003 | - | 355,003 |
| Change in Net Assets | 38,061 | (6,469) | 31,592 | 5,845 | 37,437 |
| Net Assets, Beginning of Year | 107,133 | 146,856 | 253,989 | 23,031 | 277,020 |
| Net Assets, End of Year | $ 145,194 | $ 140,387 | $ 285,581 | $ 28,876 | $ 314,457 |

See Accompanying Notes.

# LABOR AND EMPLOYMENT RELATIONS ASSOCIATION
## STATEMENT OF FUNCTIONAL EXPENSES
## FOR THE YEAR ENDED DECEMBER 31, 2022

| | | MEETINGS | | | PUBLICATIONS | | | SUPPORTING SERVICES | | |
|---|---|---|---|---|---|---|---|---|---|---|
| | General | Annual Conference | ASSA Meeting | Other Meetings | Perspectives | Research Volume | Other Publications | Management & General | Membership Development | Total |
| Compensation | $ 147,019 | $ - | $ - | $ - | $ - | $ - | $ - | $ 30,358 | $ - | $ 177,377 |
| Payroll Taxes & Fringes | 59,382 | - | - | - | - | - | - | 11,445 | - | 70,827 |
| Contract Labor | 30,519 | - | - | - | - | - | - | - | - | 30,519 |
| Depreciation | - | - | - | - | - | - | - | 1,662 | - | 1,662 |
| Insurance | - | - | - | - | - | - | - | 1,261 | - | 1,261 |
| Bank and Service Charges/Fulfillment | - | - | - | - | - | - | - | 12,733 | - | 12,733 |
| Promotion | - | - | - | - | - | - | - | - | 12,206 | 12,206 |
| Postage/Freight | - | - | - | - | 2,031 | 2,602 | - | 407 | - | 5,040 |
| Accounting/Auditing | - | - | - | - | - | - | - | 7,038 | - | 7,038 |
| Printing and Production | - | - | 614 | - | 12,229 | 7,757 | - | - | - | 20,600 |
| Services | - | 149 | - | - | 5,580 | 10,046 | - | - | - | 15,775 |
| Other Publication Costs | - | - | - | - | 78 | - | 193 | - | - | 271 |
| Meals/Receptions | - | 73 | - | - | - | - | - | - | - | 73 |
| Travel | - | 1,128 | - | 425 | - | - | - | - | - | 1,553 |
| Other Meeting Expenses | - | 65 | 255 | - | - | - | - | - | - | 320 |
| Computer Supplies/Services | - | - | - | - | - | - | - | 1,101 | - | 1,101 |
| Website | - | - | - | - | - | - | - | 133 | - | 133 |
| Office Supplies | - | - | - | - | - | - | - | 4,577 | - | 4,577 |
| Student & Member Awards | - | - | - | - | - | - | - | - | 7,587 | 7,587 |
| Fundraising Expense | - | - | - | - | - | - | - | - | 1,683 | 1,683 |
| Chapter Expenses | - | - | - | - | - | - | - | - | 6,182 | 6,182 |
| Miscellaneous Office | - | - | - | - | - | - | - | 252 | - | 252 |
| Total | $ 236,920 | $ 1,415 | $ 869 | $ 425 | $ 19,918 | $ 20,405 | $ 193 | $ 70,967 | $ 27,658 | $ 378,770 |

See Accompanying Notes.

## LABOR AND EMPLOYMENT RELATIONS ASSOCIATION
## STATEMENT OF FUNCTIONAL EXPENSES
## FOR THE YEAR ENDED DECEMBER 31, 2021

| | | MEETINGS | | | PUBLICATIONS | | | SUPPORTING SERVICES | | |
|---|---|---|---|---|---|---|---|---|---|---|
| | General | Annual Conference | ASSA Meeting | Other Meetings | Perspectives | Research Volume | Other Publications | Management & General | Membership Development | Total |
| Compensation | $ 137,638 | $ - | $ - | $ - | $ - | $ - | $ - | $ 30,943 | $ - | $ 168,581 |
| Payroll Taxes & Fringes | 60,442 | - | - | - | - | - | - | 11,464 | - | 71,906 |
| Contract Labor | 19,090 | - | - | - | - | - | - | - | - | 19,090 |
| Depreciation | - | - | - | - | - | - | - | 2,079 | - | 2,079 |
| Insurance | - | - | - | - | - | - | - | 1,339 | - | 1,339 |
| Bank and Service Charges/Fulfillment | - | - | - | - | - | - | - | 15,698 | - | 15,698 |
| Promotion | - | - | - | - | - | - | - | - | 4,595 | 4,595 |
| Postage/Freight | - | - | - | - | 1,896 | 2,301 | - | 415 | - | 4,612 |
| Accounting/Auditing | - | - | - | - | - | - | - | 6,777 | - | 6,777 |
| Printing and Production | - | - | - | - | 8,498 | 7,674 | - | - | - | 16,172 |
| Services | - | 10,000 | - | - | 7,423 | 8,108 | 595 | - | - | 26,126 |
| Other Publication Costs | - | - | - | - | 27 | - | - | - | - | 27 |
| Other Meeting Expenses | - | - | 120 | - | - | - | - | - | - | 120 |
| Computer Supplies/Services | - | - | - | - | - | - | - | 3,859 | - | 3,859 |
| Website | - | - | - | - | - | - | - | 1,845 | - | 1,845 |
| Office Supplies | - | - | - | - | - | - | - | 1,858 | - | 1,858 |
| Student & Member Awards | - | - | - | - | - | - | - | - | 4,784 | 4,784 |
| Fundraising Expense | - | - | - | - | - | - | - | - | 943 | 943 |
| Duplicating Expense | - | - | - | - | - | - | - | 1,156 | - | 1,156 |
| Chapter Expenses | - | - | - | - | - | - | - | - | 3,316 | 3,316 |
| Other Committee Expenses | - | - | - | - | - | - | - | - | 10 | 10 |
| Miscellaneous Office | - | - | - | - | - | - | - | 110 | - | 110 |
| Total | $ 217,170 | $ 10,000 | $ 120 | $ - | $ 17,844 | $ 18,083 | $ 595 | $ 77,543 | $ 13,648 | $ 355,003 |

See Accompanying Notes.

## LABOR AND EMPLOYMENT RELATIONS ASSOCIATION
## STATEMENTS OF CASH FLOWS
## FOR THE YEARS ENDED DECEMBER 31, 2022 AND 2021

| | 2022 | 2021 |
|---|---|---|
| CASH FLOWS FROM OPERATING ACTIVITIES | | |
| Change in Net Assets | $ 76,621 | $ 37,437 |
| Adjustments to Reconcile Change in Net Assets to Net Cash Provided by (Used in) Operating Activities: | | |
| Depreciation Expense | 1,662 | 2,079 |
| Net Unrealized (Gains) Losses on Investments | 6,756 | 8,749 |
| (Increase) Decrease in Operating Assets: | | |
| Accounts Receivable, Net | (367) | (1,406) |
| Prepaid Expenses | (12,593) | 2,991 |
| Inventory | (856) | 313 |
| Increase (Decrease) in Operating Liabilities: | | |
| Accounts Payable | 250 | (750) |
| Accrued Liabilities | 5,416 | (2,394) |
| Dues Collected in Advance | 8,366 | 2,931 |
| Subscriptions Collected in Advance | 4,879 | (3,581) |
| Deferred Chapter Dues | (2,350) | (2,300) |
| Other Deferred Revenue | (44,100) | 31,375 |
| Net Cash Provided by (Used in) Operating Activities | 43,684 | 75,444 |
| CASH FLOWS FROM INVESTMENT ACTIVITIES | | |
| Reinvested Interest, Dividends, and Capital Gains | (2,779) | (3,383) |
| Purchases of Property and Equipment | - | (2,643) |
| Net Cash Provided by (Used in) Investment Activities | (2,779) | (6,026) |
| CASH FLOWS FROM FINANCING ACTIVITIES | | |
| None | - | - |
| Net Cash Provided by (Used in) Financing Activities | - | - |
| Net Increase (Decrease) in Cash and Cash Equivalents | 40,905 | 69,418 |
| Beginning Cash and Cash Equivalents | 394,971 | 325,553 |
| Ending Cash and Cash Equivalents | $ 435,876 | $ 394,971 |

See Accompanying Notes.

## NOTE 1 – NATURE OF ACTIVITIES AND SIGNIFICANT ACCOUNTING POLICIES

### Organization and Nature of Activities

The Labor and Employment Relations Association (the Organization) was founded in 1947 to encourage research in all aspects of the field of labor, employment, and the workplace. It is a nonprofit scholarly association of academic, labor, business, and neutral communities committed to the full discussion and exchange of ideas between and among its broad constituencies through meetings, publications, and its various electronic listservs and websites. The Organization's National Office is located in Champaign, Illinois and serves the association by planning conferences and meetings and publishing the various researches of its members. The main sources of support and revenues for the Organization are contributions, membership dues and subscriptions, and meeting income.

### Subsequent Events

Subsequent events have been evaluated through May 30, 2023, which is the date the financial statements were available to be issued. There were no subsequent events requiring recognition or disclosure in the financial statements.

### Accrual Basis of Reporting

The Organization has chosen to report on the accrual basis of accounting. Accordingly, revenue is recognized when earned and expenses are recognized when incurred in conformity with accounting principles generally accepted in the United States of America (GAAP). The financial statements are presented in accordance with Financial Accounting Standards Board (FASB) Accounting Standards Codification (ASC) 958 dated August 2016, and the provisions of the American Institute of Certified Public Accountants (AICPA) "Audit and Accounting Guide for Not-for-Profit Organizations" (the "Guide").

### Use of Estimates in Preparing Financial Statements

The preparation of financial statements in conformity with GAAP requires management to make estimates and assumptions that affect certain reported amounts and disclosures. Accordingly, actual results could differ from those estimates.

### Cash and Cash Equivalents

For purposes of reporting cash flows, cash and cash equivalents include all cash and highly liquid investments acquired with an original maturity date of three months or less. As of December 31, 2022 and 2021, $63,454 and $65,796 of cash and cash equivalents are designated by the board and are subject to board-imposed stipulations. As of December 31, 2022 and 2021, $34,904 and $28,876 of cash and cash equivalents are temporarily restricted and are subject to donor stipulations.

**Investments**

Investments consist of intermediate term bond funds, Bayer AG (BAYRY) stock, and money market funds, which are carried at the fair value of the underlying assets. Net appreciation (depreciation) in the fair value of investments, which consists of the realized gains or losses and the unrealized appreciation (depreciation) on those investments, is presented in the statement of activities in accordance with donor restrictions as investment income. Investment income is presented net of investment fees. The average cost method is primarily used to determine the basis for computing realized gains or losses.

**Accounts Receivable**

Accounts receivable are recorded primarily for outstanding invoices for membership dues and subscriptions. An allowance for doubtful accounts is based on an analysis of expected collection rates determined from experience. The Organization had an allowance for doubtful accounts of $0 and $0 at December 31, 2022 and 2021, respectively.

**Inventory**

The Organization's inventory of directories, research volumes, proceedings, and perspective magazines is carried at the lower of cost and market value. Cost is determined on the basis of first in – first out.

**Property and Equipment**

Property and equipment expenditures in excess of $1,000 are capitalized at cost. Donated property and equipment are capitalized at estimated cost or fair market value at the time of donation. Depreciation of the assets is computed using the straight-line method over their estimated useful lives. The range of estimated useful lives by type of asset is as follows:

| | |
|---|---|
| Furniture and Equipment | 3 - 7 years |

**Net Assets**

Net assets of the Organization and changes therein are classified and reported as follows:

Net Assets without Donor Restrictions - Net assets that are not subject to donor-imposed restrictions and may be expended for any purpose in performing the primary objectives of the Organization. The Organization's board may designate assets without restrictions for specific operational purposes from time to time.

Net Assets with Donor Restrictions - Net assets subject to stipulations imposed by donors, and grantors. Some donor restrictions are temporary in nature; those restrictions will be met by actions of the Organization or by the passage of time. Other donor restrictions are perpetual in nature, where by the donor has stipulated the funds be maintained in perpetuity. As of December 31, 2022 and 2021, the Organization had no donor restrictions that were perpetual in nature.

## Contributions

Unconditional contributions are recognized when pledged and recorded as net assets without donor restrictions or net assets with donor restrictions, depending on the existence and/or nature of any donor-imposed restrictions. Conditional promises to give are recognized when the conditions on which they depend are substantially met. Gifts of cash and other assets are reported with donor restricted support if they are received with donor stipulations that limit the use of the donated assets.

When a restriction expires, that is, when a stipulated time restriction ends or a purpose restriction is accomplished, net assets with donor restrictions are reclassified to net assets without donor restrictions and reported in the statement of activities as net assets released from restrictions. Donor-restricted contributions whose restrictions are met in the same reporting period are reported as net assets without donor restriction support. Contributions restricted for the acquisition of land, buildings, and equipment are reported as net assets without donor restriction upon acquisition of the assets and the assets are placed in service.

## Contributed Services and Goods

Contributed services are reported as contribution revenue and as assets or expenses only if the services create or enhance a non-financial asset (for example property and equipment) or:

- Would typically need to be purchased by the Organization if the services had not been provided by contribution.
- Require specialized skills.
- Are provided by individuals with those skills (such as accounting, financial, construction, educational, electrical, legal, medical, and other services provided by accountants, investments advisers, contractors, teachers, electricians, lawyers, doctors, and other professional and craftspeople).

For the years ended December 31, 2022 and 2021, the value of contributed services meeting the defined requirements for recognition in the financial statements as outlined above were not material and have not been recorded on the financial statements. The Organization does receive free office space from the University of Illinois. However, the value for the use of this office space has been determined to be immaterial.

## Income Tax Status

The Organization is a nonprofit association that is exempt from federal income tax under Section 501(c)(3) of the Internal Revenue Code. In addition, the Internal Revenue Service has determined that the Organization is not a private foundation as defined in Section 509(a)(1) and Section 170(b)(1)(A)(vi) of the Code.

The Organization has evaluated its exposure resulting from uncertain income tax position and determined the exposure is not material to the financial statements. In addition, the Organization is not aware of any tax position for which a significant change is reasonably possible within the next 12 months. Therefore, these financial statements do not include a liability for uncertain tax positions. Upon recognition of a liability for an uncertain tax position, the Organization would recognize interest expense and penalties in operating expenses.

The Organization files information tax returns in the U.S. federal jurisdiction and the state of Illinois. Its federal and Illinois information tax returns prior to calendar year 2019 are closed. The Organization does not have any tax returns currently under examination by either the Internal Revenue Service (IRS) or any U.S. state jurisdiction.

**Membership Dues and Advance Subscriptions Collected**

Membership dues and subscriptions are assessed and recognized as revenue based on the life of the dues or subscription.

**Functional Allocation of Expenses**

The costs of providing the various programs and other activities have been summarized on a functional basis in the statements of activities. The statements of functional expenses present the natural classification detail of expenses by function. Accordingly, certain costs have been allocated among the programs and supporting services benefited.

Expenses which are easily and directly associated with a particular program or supporting service are charged directly to that functional area. Compensation and payroll taxes & fringes have been allocated on the basis of estimates of time and effort.

NOTE 2 – ARRANGEMENTS WITH THE UNIVERSITY OF ILLINOIS

The Organization moved its offices to the University of Illinois at the end of 1999. Under an arrangement with the University, the employees of the Organization are employed by the University. The employees' pension and benefits are part of the University's plans. The Organization then reimburses the University monthly for the cost of its employees.

The University of Illinois holds some cash for the Organization. These "claim on cash" balances were $49,189 and $20,584 as of December 31, 2022 and 2021, respectively. These balances have been included with cash and cash equivalents.

NOTE 3 – CONCENTRATIONS OF CREDIT RISK

Financial instruments that potentially subject the Organization to credit risk consist principally of checking accounts, money markets accounts, and certificates of deposits at financial institutions. However, management continuously monitors the Organization's balances at financial institutions. The balances at each bank as of December 31, 2022 and 2021 were insured by the Federal Deposit Insurance Corporation (FDIC) up to $250,000. As of December 31, 2022 and 2021, $105,713 and $84,767, respectively, of the bank balance was not insured by FDIC. The "claim on cash" held with the University of Illinois as stated in Note 2 is not insured since it is not held in a financial institution. In addition, the Organization had undeposited funds of $600 and $5,834 at December 31, 2022 and 2021, respectively. Since these funds were not yet deposited with a financial institution it was not covered by FDIC.

NOTE 4 – INVESTMENTS

FASB ASC 820, *Fair Value Measurements*, establishes a framework for measuring fair value. That framework provides a fair value hierarchy that prioritizes the inputs to valuation techniques used to measure fair value. The hierarchy gives the highest priority to unadjusted quoted prices in active markets for identical assets or liabilities (Level 1 measurements) and the lowest priority to unobservable inputs (Level 3 measurements). The three levels of the fair value hierarchy under FASB ASC 820 are described below:

> Level 1 – Inputs to the valuation methodology are based on unadjusted quoted prices for identical assets or liabilities in active markets that the Organization has the ability to access.
>
> Level 2 – Inputs to the valuation methodology include quoted prices for similar assets or liabilities in active markets, quoted prices for identical or similar assets or liabilities in inactive markets, inputs other than quoted prices that are observable for the asset or liability, and inputs that are derived principally from or corroborated by observable market data by correlation or other means. If the asset or liability has a specified (contractual) term, the Level 2 input must be observable for substantially the full term of the asset or liability.
>
> Level 3 – Inputs to the valuation methodology are unobservable and significant to the fair value measurement.

The asset's and liability's fair value measurement level within the fair value hierarchy is based on the lowest level of any input that is significant to the fair value measurement. Valuation techniques used need to maximize the use of observable inputs and minimize the use of unobservable inputs.

The following is a description of the valuation methodologies used for assets measured at fair value. There have been no changes in the methodologies used at December 31, 2022 and 2021.

| | |
|---|---|
| *Intermediate Term Bond Funds:* | Valued at the net asset value (NAV) of shares held by the Organization at year-end based on readily determinable fair values, which are published daily and are the basis for current transactions. |
| *Bayer AG (BAYRY) Stock:* | Valued at the stated share price at year-end. |
| *Money Market Funds*: | Valued at cost plus accrued interest. |

The preceding methods described may produce a fair value calculation that may not be indicative of net realizable value or reflective of future fair values. Furthermore, although the Organization believes its valuation methods are appropriate and consistent with other market participants, the use of different methodologies or assumptions to determine the fair value of certain financial instruments could result in different fair value measurements at the reporting date.

The following table sets forth by level, within the fair value hierarchy, the Organization's assets at fair value as of December 31, 2022 and 2021:

*Assets at Fair Value as of December 31, 2022*

| | Level 1 | Level 2 | Level 3 | Total |
|---|---|---|---|---|
| *Without Donor Restrictions* | | | | |
| Bayer AG (BAYRY) | $ 10,288 | - | - | $ 10,288 |
| Money Market Funds | 1,055 | - | - | 1,055 |
| *Designated by the Board* | | | | |
| Intermediate Term Bond Funds | 25,263 | - | - | 25,263 |
| Bayer AG (BAYRY) | 41,152 | - | - | 41,152 |
| Money Market Funds | 4,220 | - | - | 4,220 |
| Total | $ 81,978 | - | - | $ 81,978 |

*Assets at Fair Value as of December 31, 2021*

| | Level 1 | Level 2 | Level 3 | Total |
|---|---|---|---|---|
| *Without Donor Restrictions* | | | | |
| Bayer AG (BAYRY) | $ 10,608 | - | - | $ 10,608 |
| Money Market Funds | 756 | - | - | 756 |
| *Designated by the Board* | | | | |
| Intermediate Term Bond Funds | 29,129 | - | - | 29,129 |
| Bayer AG (BAYRY) | 42,432 | - | - | 42,432 |
| Money Market Funds | 3,030 | - | - | 3,030 |
| Total | $ 85,955 | - | - | $ 85,955 |

Investment income on the statements of activities for the years ended December 31, 2022 and 2021 are summarized as follows:

| | 2022 | 2021 |
|---|---|---|
| Dividends | $ 2,772 | $ 2,966 |
| Capital Gains Distributions | 7 | 417 |
| Net Unrealized Gains (Losses) | (6,756) | (8,749) |
| Total Investment Income | $ (3,977) | $ (5,366) |

This investment income, including unrealized gains and losses, are being reported on the statements of activities as investment income (loss).

## NOTE 5 – PROPERTY AND EQUIPMENT

As of December 31, 2022, Property and Equipment consists of:

| | |
|---|---|
| Furniture and Equipment | $ 21,303 |
| Less: Accumulated Depreciation | (19,673) |
| Property and Equipment, Net | $ 1,630 |

As of December 31, 2021, Property and Equipment consists of:

| | |
|---|---|
| Furniture and Equipment | $ 21,303 |
| Less: Accumulated Depreciation | (18,011) |
| Property and Equipment, Net | $ 3,292 |

Depreciation expense for the years ended December 31, 2022 and 2021 was $1,662 and $2,079, respectively.

## NOTE 6 – NET ASSETS - BOARD DESIGNATED

### Susan C. Eaton Scholar-Practitioner Memorial Fund

The Organization set up a memorial fund in honor of an author of a 1998 "Perspectives on Work" article who died on December 30, 2003. At its June 1, 2004 meeting, the Organization approved the establishment of an annual Susan C. Eaton Scholar-Practitioner award and grant to be paid to one or more qualified scholar researchers in even-numbered years or practitioners in odd-numbered years doing research in the labor and employment relations or related field. The Organization's Executive Board directed that 10% of any gifts received each year are undesignated and can be used for administrative expenses. As of December 31, 2022 and 2021, these designated funds totaled $16,779 and $20,779, respectively.

### Kochan-Sleigh Best Dissertation Award Fund

On March 27, 2006, the Organization set up a designated fund with contributions from a member and matching contributions from General Electric. The fund is designated for a minimum of fifteen years and may be used to pay for the $1,000 annual best dissertation award and plaque. After the fifteen-year period, the fund may be continued or the amount remaining in the fund may be undesignated and become available to the Organization for unrestricted purposes. The Organization's Executive Board directed that 10% of any gifts received each year are undesignated and can be used for administrative expenses. As of December 31, 2022 and 2021, these designated funds totaled $35,319 and $36,439, respectively.

**Jim Scoville Award Fund**

On January 9, 2009, the Organization set up a designated fund with a $10,000 contribution from the University of Minnesota's Industrial Relations Center to honor a member and retiring professor. The fund was established to pay an annual award of $500 and a plaque for best paper on international and comparative employment issues. The Organization's Executive Board directed that 10% of any gifts received each year are undesignated and can be used for administrative expenses. As of December 31, 2022 and 2021, these designated funds totaled $2,176 and $2,796, respectively.

**John T. Dunlop Public Policy Fund**

On May 8, 2010, the Organization established a designated fund to subsidize a named John T. Dunlop Public Policy Session at future National Policy Forums and/or at the Organization annual meetings. The Organization's Executive Board directed that 10% of any gifts received each year are undesignated and can be used for administrative expenses. This fund will be so named for five or ten years, at which time the Organization's Executive Board will review the fund to determine whether to continue or modify the fund. As of December 31, 2022 and 2021, these designated funds totaled $16,503 and $18,326, respectively.

**Gladys and Walter Gershenfeld Diversity & Inclusion Fund**

On May 8, 2010, the Organization established a designated fund to honor the memory of long-time members and past presidents Walter and Gladys Gershenfeld for the purpose of supporting the Organization's electronic and print publications. The fund will be so named for ten years, at which time the Organization will review the fund's purpose and uses, and whether to continue or modify the fund. The Organization's Executive Board directed that 10% of any gifts received each year are undesignated and can be used for administrative expenses. As of December 31, 2022 and 2021, these designated funds totaled $17,939 and $16,585, respectively.

**Quasi-Endowment Fund**

During 2019, the Organization established a quasi-endowment fund for the purpose of supporting the Organization's general operations. The Organization's Executive Board directed that 20% of any gifts received each year are undesignated and can be used for administrative expenses. Funds designated by the Board of Directors to function as an endowment are voluntary and may be reversed by the Board of Directors at any time. Accordingly, they are reported as part of the net assets without donor restrictions. The Organization's Board designated endowment net assets exist as a permanent investment pool to enhance and sustain the operations of the Organization so that it can continue to be involved in the full discussion and exchange of ideas between and among its broad constituencies through meetings, publications, and its various electronic listservs and websites. As of December 31, 2022 and 2021, these designated funds totaled $45,373 and $45,462, respectively.

## NOTE 7 – NET ASSETS WITH DONOR RESTRICTIONS

Net assets with donor restrictions at December 31, 2022 and 2021 are restricted for the following purposes or period:

| | 2022 | 2021 |
|---|---|---|
| Subject to expenditures for specified purpose: | | |
| Administer the Business of the Deans, Chairs, and Directors Committee (DCDC) | $ 8,460 | $ 8,320 |
| Administer the PhD Student Consortium | 26,444 | 20,556 |
| Total Net Assets with Donor Restrictions | $ 34,904 | $ 28,876 |

Net assets were released from donor restrictions by incurring expenses satisfying the restricted purposes specified by the donors for the years ended December 31, 2022 and 2021:

| | 2022 | 2021 |
|---|---|---|
| Administer the Business of the Deans, Chairs, and Directors Committee (DCDC) | $ 410 | $ 350 |
| | $ 410 | $ 350 |

## NOTE 8 – ENDOWMENT

Board designated net assets at December 31, 2022 and 2021 consist of an endowment fund established in 2019 for the purpose of supporting operations of the Organization. The Organization has adopted the provisions of FASB ASC 958-205-45, *Not-for-Profit Entities,* including *Net Asset Classification of Funds Subject to an Enacted Version of the Uniform Prudent Management of Institutional Funds Act, and Enhanced Disclosures for All Endowment Fund.* ASC 958 provides guidance on the net asset classification of donor restricted endowment funds for a non-for-profit organization that is subject to an enacted version of the State Prudent Management of Institutional Funds Act (SPMIFA) and also requires disclosures about endowment funds, both donor restricted and board designated endowment funds.

*Interpretation of Relevant Law*

The Organization has interpreted SPMIFA as requiring the preservation of the fair value of the original gift as of the gift date of the donor-restricted endowment funds absent explicit donor stipulations to the contrary. As a result of this interpretation, the Organization classifies as permanently restricted net assets (a) the original value of gifts donated to the permanent endowment, (b) the original value of subsequent gifts to the permanent endowment, and (c) accumulations to the permanent endowment made in accordance with the direction of the applicable donor gift instrument at the time the accumulation is added to the fund. The remaining portion of the donor-restricted endowment fund that is not classified as permanently restricted net assets is classified as net assets with donor restrictions - restricted for specified purpose until those amounts are appropriated for expenditures by the Organization in a manner consistent with the standard of prudence prescribed by SPMIFA.

Additionally, in accordance with SPMIFA, the Organization considers the following factors in making a determination to appropriate or accumulate donor-restricted endowment funds:

(1) The duration and preservation of the fund
(2) The purpose of the Organization and the donor-restricted endowment fund
(3) General economic conditions
(4) The possible effect of inflation or deflation
(5) The expected total return from income and the appreciation of investments
(6) Other resources of the Organization
(7) The investment policies of the Organization.

*Endowment Net Asset Composition and Changes in Net Assets*
The composition of endowment net assets and the changes in endowment net assets as of December 31, 2022 and 2021 are as follows:

| | Without Donor Restrictions | |
|---|---|---|
| | 2022 | 2021 |
| Endowment Net Assets, Beginning of Year | $ 45,462 | $ 49,190 |
| Member Contributions | - | - |
| Investment Income | (89) | (3,728) |
| Endowment Net Assets, End of Year | $ 45,373 | $ 45,462 |

*Return Objectives and Risk Parameters*
The primary objectives in the investment management for the portfolio assets is to provide ongoing stability and income for operations, preserving gift assets without subjecting them to substantial risk, and providing additional real growth through new gifts. The mix of investments in the portfolio is examined periodically.

*Spending Policy*
Spending from the fund should not exceed two to five percent annually depending on growth, general economic conditions, and organizational needs. The spending policy should be reviewed every five to ten years.

## NOTE 9 – LIQUIDITY AND AVAILABILITY OF FINANCIAL ASSETS

The following reflects the Organization's financial assets as of the statements of financial position date, reduced by amounts not available for general use because of contractual or donor-imposed restrictions within one year of the statement of financial position date. Amounts available include donor restricted amounts that are available for general expenditure in the following year.

| | 2022 | 2021 |
|---|---|---|
| Fiscal Assets at Year-End | | |
| Cash and Cash Equivalents | $ 435,876 | $ 394,971 |
| Investments | 81,978 | 85,955 |
| Accounts Receivable, Net | 1,773 | 1,406 |
| Less Contractual or Donor-Imposed Restrictions | | |
| Board Designated Funds | (134,089) | (140,387) |
| Donor Restrictions for Specific Purposes | (34,904) | (28,876) |
| Financial Assets Available to Meet Cash Needs for General Expenditure Within One Year | $ 350,634 | $ 313,069 |

As part of the Organization's liquidity management, it has a policy to structure its financial assets to be available as its general expenditures, liabilities, and other obligations come due. Excess cash is generally held in checking accounts, money markets accounts, and certificates of deposits until it is required for operational use.

NOTE 10 – COMMITMENTS

The Organization signed a contract with the MotorCity Casino Hotel (the Hotel) in Detroit on November 7, 2019 to hold the 73rd Annual Meeting from June 1st through the 7th, 2021. Due to COVID-19, the Annual Meeting went virtual in 2020 and stayed virtual for 2021 and 2022. As a result, on September 10, 2020, the Organization did an addendum to the contract with the Hotel in Detroit to change the Annual Meeting date to May 31st through June 6th, 2022. On February 15, 2022, the Organization did another addendum to the contract with the Hotel in Detroit to change the Annual Meeting date to 2023, with a cut-off date of April 30, 2023. All other conditions of the contract including concessions and meeting space apply as originally negotiated. The Organization signed a contract with the New York Hilton Midtown in New York on December 21, 2022 to hold the 76th Annual Meeting from June 26th through July 1st, 2024. If these contracts were cancelled at December 31, 2022, the Organization would have owed $121,311 to the Hotel in Detroit, and $184,954 to the New York Hilton Midtown in New York for a total of $306,265. If these contracts were cancelled at December 31, 2021, the Organization would have owed $40,437 to the Hotel in Detroit.

NOTE 11 – PENSION PLAN AND RETIREMENT CONTRIBUTIONS

The University of Illinois (the University) contributes to the State Universities Retirement System of Illinois (SURS), a cost-sharing multiple-employer defined benefit plan with a special funding situation whereby the State of Illinois (the State) makes substantially all actuarially determined required contributions on behalf of the participating employers. SURS was established July 21, 1941 to provide retirement annuities and other benefits for staff members and employees of state universities, certain affiliated organizations, and certain other state educational and scientific agencies and for survivors, dependents, and other beneficiaries of such employees. SURS is considered a component unit of the State of Illinois' financial reporting entity and is included in the State's financial reports as a pension trust fund. SURS is governed by

Section 5/15, Chapter 40 of the *Illinois Compiled Statutes*. SURS issues a publicly available financial report that includes financial statements and required supplementary information. That report may be obtained by accessing the website at www.SURS.org.

*Benefits Provided.* A traditional benefit plan was established in 1941. Public Act 90-0448, enacted effective January 1, 1998, established an alternative defined benefit program known as the portable benefit package. The traditional and portable plan Tier 1 refers to members who began participation prior to January 1, 2011. Public Act 96-0889 revised the traditional and portable benefit plans for members who begin participation on or after January 1, 2011, and who do not have other eligible Illinois reciprocal system services. The revised plan is referred to as Tier 2. New employees are allowed six months after their date of hire to make an irrevocable election. A summary of the benefit provisions can be found in the SURS comprehensive annual financial report (CAFR) notes to the financial statements.

*Contributions.* The State of Illinois is primarily responsible for funding SURS on behalf of the individual employers at an actuarially determined amount. Public Act 88-0593 provides a statutory funding plan consisting of two parts: (i) a ramp-up period from 1996 to 2010 and (ii) a period of contributions equal to a level percentage of the payroll of active members of SURS to reach 90 percent of the total actuarial accrued liability by the end of fiscal year 2045. Employer contributions from "trust, federal, and other funds" are provided under Section 15-155(b) of the Illinois Pension Code and require employers to pay contributions which are sufficient to cover the accruing normal costs on behalf of applicable employees. The employer normal cost was 12.70 percent of employee payroll during the period of July 1, 2020 to June 30, 2021. The employer normal cost was 12.32 percent of employee payroll during the period of July 1, 2021 to June 30, 2022. The employer normal cost was 12.83 percent of employee payroll during the period of July 1, 2022 to June 30, 2023. The normal cost is equal to the value of current year's pension benefit and does not include any allocation for the past unfunded liability or interest on the unfunded liability. Plan members are required to contribute 8.0 percent of their annual covered salary. The contribution requirements of plan members and employers are established and may be amended by the Illinois General Assembly. During the years ended December 31, 2022 and 2021, the Organization reimbursed the University of Illinois $22,349 and $20,280, respectively, for the employer's payments to SURS for eligible employees.

In addition to providing pension benefits, the State of Illinois provides certain health, dental and life insurance benefits to annuitants. This includes annuitants of the Organization. Substantially all State employees, including the Organization's employees, may become eligible for postemployment benefits if they eventually become annuitants. Health and dental benefits include basic benefits for annuitants under the State's self-insurance plan and insurance contracts currently in force. Life insurance benefits for annuitants under age 60 are equal to their annual salary at the time of retirement; life insurance benefits for annuitants age 60 or older are limited to $5,000 per annuitant. Currently, the State does not segregate payments made to annuitants from those made to current employees for health, dental, or life insurance benefits. These costs are funded by the State and are not an obligation of the Organization.

Employees of the Organization may also elect to participate in several tax deferred annuity plans and defined contribution plans. These are single employer plans under which benefits are provided to participating employees through contracts issued to each individual. Participation and the level of employee contributions are voluntary. The Organization is not required to make contributions.

*LERA gratefully acknowledges the continuing support of its*
*2022–23 Major Sponsors and Organizational Members*

**Major Sustaining Sponsors**

International Union, UAW

UAW-Ford Joint Trust

Kaiser Permanente and the Alliance of Health Care Unions

**Major University Sponsors**

Cornell University, School of Industrial and Labor Relations

Florida International University, College of Business

Massachusetts Institute of Technology, Institute for Work and Employment Research

Michigan State University, School of Human Resources and Labor Relations

Pennsylvania State University, School of Labor and Employment Relations

Princeton University, Industrial Relations Section

Rutgers University, School of Management and Labor Relations

University of Illinois at Urbana-Champaign, School of Labor & Employment Relations

**Organizational Members**

American Federation of Teachers, AFL-CIO

American Train Dispatchers Association

Boston University, Questrom School of Business

Centre for Employment and Labour Relations Law / Melbourne Law School

Cornell University, Scheinman Institute on Conflict Resolution

Cornell University, School of Industrial and Labor Relations

Federal Mediation and Conciliation Service

Florida International University, College of Business

IAMAW

International Union, UAW

Kaiser Permanente and the Alliance of Health Care Unions

Keystone Research Center

Massachusetts Institute of Technology, Sloan School of Management

Michigan State University, School of Human Resources and Labor Relations

New Jersey School Boards Association

Pennsylvania State University, School of Labor and Employment Relations

Portland General Electric

Portland State University

Princeton University, Industrial Relations Section

Rutgers School of Management and Labor Relations

Taiwan Institute of Labor, Occupational Safety and Health

UAW-Ford Joint Trust

UNAC/UHCP

United Mine Workers of America

U.S. Department of Agriculture

University of California, Los Angeles, Anderson School of Management

University of California-Berkeley, Center for Labor Research and Education

University of California-Berkeley, Institute for Research on Labor and Employment

University of Illinois at Urbana-Champaign, School of Labor and Employment Relations
University of Minnesota, Center for Human Resources and Labor Studies
University of Toronto, Centre for Industrial Relations and Human Resources
University of Wisconsin School for Workers
Wayne State University, Labor@Wayne
W. E. Upjohn Institute for Employment Research

For more information about Organizational Membership with the LERA, please visit the LERA website: https://lera.memberclicks.net/organizational-membership.

# ALPHABETICAL LIST OF AUTHORS